Dominick E. Fazarro, Christie M. Sayes, Walt Trybula, Jitendra S. Tate, Cra

Nano-Safety

Also of Interest

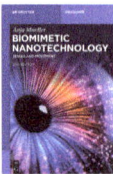

Biomimetic Nanotechnology
Senses and Movement
Anja Mueller, 2023
ISBN 978-3-11-077918-9, e-ISBN 978-3-11-077919-6

Microbial Nanotechnology
Advances in Agriculture, Industry and Health Sectors
Edited by: Vikas Kumar and Manoj Singh, 2022
ISBN 978-3-11-037914-3, e-ISBN 978-3-11-037916-7

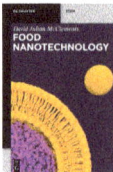

Food Nanotechnology
David Julian McClements, 2022
ISBN 978-3-11-078842-6, e-ISBN 978-3-11-078845-7

Corrosion Prevention Nanoscience
Nanoengineering Materials and Technologies
Edited by: Berdimurodov Elyor Tukhlievich and Chandrabhan Verma, 2023
ISBN 978-3-11-107009-4, e-ISBN 978-3-11-107175-6

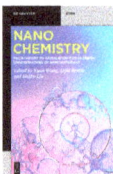

Nanochemistry
From Theory to Application for In-Depth Understanding of Nanomaterials
Edited by: Xuan Wang, Sajid Bashir and Jingbo Liu, 2022
ISBN 978-3-11-073985-5, e-ISBN 978-3-11-073987-9

Nano-Safety

What We Need to Know to Protect Workers

Edited by
Dominick E. Fazarro, Christie M. Sayes, Walt Trybula,
Jitendra S. Tate, Craig Hanks

2nd edition

DE GRUYTER

Editors

Dr. Dominick E. Fazarro
The University of Texas at Tyler
3900 University Blvd
Tyler, TX 75799
USA
dfazarro@uttyler.edu

Dr. Christie M. Sayes
Baylor Sciences Building
#97266 One Bear Place
Waco, TX 76798-7266
USA

Ph. D. Walt Trybula
The Trybula Foundation
4621A Pinehurst Dr. South
Austin, TX 78747
USA
w.trybula@tryb.org

Dr. Jitendra S. Tate
Texas State University
6608 Aden Ln
Austin, TX 78739
USA

Dr. Craig Hanks
Texas State University
105 Pickard St. DERR 201
San Marcos, TX 7866
USA
craig.hanks@txstate.edu

ISBN 978-3-11-078182-3
e-ISBN (PDF) 978-3-11-078183-0
e-ISBN (EPUB) 978-3-11-078195-3

Library of Congress Control Number: 2023944101

Bibliographic information published by the Deutsche Nationalbibliothek
The Deutsche Nationalbibliothek lists this publication in the Deutsche Nationalbibliografie;
detailed bibliographic data are available on the Internet at http://dnb.dnb.de.

© 2024 Walter de Gruyter GmbH, Berlin/Boston
Cover image: Olemedia / E+ / Getty Images
Typesetting: VTeX UAB, Lithuania
Printing and binding: CPI books GmbH, Leck

www.degruyter.com

Foreword

Nanotechnology, or more correctly, the application of nanoscale science and engineering to material science, has been the focus of a major global R&D investment over the past 20 years. In the United Sates alone, the public investment has been over $ 23 Billion since 2000, with an equal investment in the private sector. To help organize its efforts in this vital area, the U.S. created the National Nanotechnology Initiative (NNI) in 2000 to help coordinate the activities of 20 federal agencies engaged in one or more aspects of the national strategy to promote the technology. Developing the scientific knowledge needed to address questions and concerns about the possible environmental, health, and safety impact of this new technology was recognized as a critical need early in the program. The fact that developing the science needed to support safe and responsible development of nanotechnology was listed in the first US Nanotechnology Strategic plan was very exciting to the practitioners, especially the occupational health and safety community. Delivering on the expectations set with that first plan have been a challenge, but good progress is being made.

As a PhD chemist, an occupational health professional, and a Board Certified Industrial Hygienist, I am excited about the diversity and promise of the output of nanotechnology, yet I am aware of the difficult science needed to answer key questions about the possible implications on human health and the environment. Many of the traditional precepts of safety and health have to be reexamined, if not redesigned to examine the potential impact on human health that this new and promising form of material science represents. If we do not meet the challenge of understanding the possible hazards, evaluating and quantifying any potential risk and, most importantly, developing solutions to manage any risk and support responsible development of this amazing technology, it is possible that, ultimately, we may not realize the many great benefits to society that are currently being explored.

I have had the privilege of managing the Nanotechnology Research Center at the National Institute for Occupational Safety and Health, NIOSH, for the past 10 years. My duties at NIOSH, coupled with the 16 years I spent in the private sector, have given me a unique opportunity to interact with some of the best scientists in every aspect of the current challenge. I am honored to work with some of the best minds in occupational safety and health and to interact with the thought leaders in nanoscale science and engineering as I promote a message of practical yet effective methods to evaluate and manage potential risks of nanotechnology. The health and safety and scientific communities have been very receptive to taking a precautionary approach with the technology and materials while it is still evolving, recognizing that high-quality science is being conducted in a concurrent fashion to develop very detailed answers to key health and safety questions. One of most encouraging things I have seen throughout this challenge is that we have a highly significant number of interested and invested parties engaged in asking and addressing the health, safety, and environment questions out in advance of the technology. The old paradigm is reacting to incidents or disasters, and then formulating solutions

https://doi.org/10.1515/9783110781830-201

after-the-fact. The former may well be one of the first broad-based proactive approach being taken to address key issues of a technology as it is being developed. This strategy would not be possible without the active participation, collaboration, and support of scientists engaged in the actual development of the technology. This is why I am so pleased to see the development of Nano-Safety: What we need to know to protect workers. Dr. Dominick Fazarro and colleagues has done a wonderful job of gathering up experts from a variety of technical areas to contribute their expert knowledge to this work. This reference takes a very pragmatic view of what is needed to educate students, researchers, faculty, and safety practitioners on the basic knowledge needed to approach this very complex issue.

I am very happy to see the variety of topics this book develops, including areas usually not seen in "traditional" safety reference, such as Ethics, Communication, Behavior, Reliability of Information, and evaluating competence of workers and safety professionals. I believe the reader will find this book to be a valuable addition to their resource library and, most importantly, a valuable tool for developing an effective strategy to keep students, researchers, faculty, engineers, line operators, and all other workers involved in the responsible development of nanotechnology health and safety.

Charles (Chuck) Geraci, Jr, PhD, CIH, FAIHA
Associate Director, National Institute for Occupational Safety and Health

This Forward from the original publication is still pertinent. If anything, the guidance provided in this volume is more important than ever.

Walt Trybula, Editor

Preface 2nd Ed

This book is the update of the initial volume, which was the culmination of more than eight years of effort. In 2006, the Nanomaterials Application Center at Texas State University was involved in evaluating nanotechnology for commercialization. One of the items that was raised was "safety." At that time, there were concerns about the impact of nanomaterials on both people and the environment. There was a significant amount of work being done on toxicity, but it was focused on individual cases. A white paper was produced that listed elements required to ensure the safety and proper handling of nanotechnology.

An award by a government agency to develop any aspects of safety for nanotechnology came from Occupational Safety and Health Administration (OSHA) as a Susan Harwood award. It was awarded to Kristen Kulinowski at Rice University to develop an eight-hour course for practitioners. Texas State supported the Rice contract, with Walt Trybula as principle investigator (PI). The resultant effort was tested at a number of national conferences. The response from course participants indicated that the eight-hour course was preferred over a four-hour course. That course is available through OSHA.

Additional efforts followed from that contract at Texas State University. Several proposals were submitted to the National Science Foundation (NSF) over a three-year period. There was an NSF award made to Texas State in 2013, with Jitendra Tate as PI and Dominick Fazarro of UT Tyler and Craig Hanks of Texas State as co-PIs. The award was to develop nanotechnology safety education courses. One course was to be at an introductory level, and the other course was to include more advanced topics. Each course consisted of nine independent modules to be used for standalone insertion into existing courses. During the development of the contents of the courses, it was apparent that there was a need to consider ethical concerns in the courses. Judgment calls are required when working with materials with unknown properties, so the courses needed to include some basis for making decisions. Those decisions need to look at the impact on people and the environment. Inclusion of the topic of ethics was considered a necessity by the team. The material was developed and tested with students at Texas State University and UT Tyler. UT Tyler offered complete courses, whereas Texas State courses were in modular form, enabling insertion of individual modules into a number of existing courses. The feedback from students and NSF evaluators provided the opportunity to tune the courses. This material is available from NSF. For this work, the team was honored by the National Academy of Engineering as having one of the 25 exemplary programs in engineering ethics.

After developing the courses, with successful response to the material, it was time to develop an educational book to be used to supplement the work on nanotechnology education. There was an outreach to leaders in the field to contribute chapters to the book. The editors are professionals who have contributed to this work.

In addition to the technologies developed over the five years since the original publication, this book has had two years in updating and proofing in order to ready it for

https://doi.org/10.1515/9783110781830-202

publication. The content is structured to provide a logical flow from one topic to the next. The first chapter, *"The World of Nanotechnology"* by Barbara Foster, starts by examining the world of nanotechnology and safety. Nanotechnology is not the first technological development that might contain significant risks to both people and the environment. Proper planning enables the development of controls to ensure the safety of workers involved. In this first chapter, Barbara Forster discusses previous development of new technology and mentions how some of the controls were established. Chapter 2, *"The World of Engineering Nanomaterials"* by Eylem Asmatulu, provides an overview of how the understanding and application of nanomaterials have evolved. Chapter 3, *"The Importance of Safety for Manufacturing Nanomaterials"* by W. S. Khan and R. Asmatulu, presents the involvement of safety with manufacturing and usage of nanomaterials. Chapter 4, *"Safety Approaches to Handling Engineered Nanomaterials"* by Jitendre S. Tate and Roger A. Hernandez, covers engineered nanomaterials and some of the possible sources of health hazards. Chapter 5, *"Certification: Validating Workers' Competence in Nano-safety"* by Christie M. Sayes, James Y. Liu, and Matthew Gibb, addresses the need for training in safety and certification to ensure that safety is dominant in nanotechnology efforts. Chapter 6, *"Understanding the Implications of Nanomaterial Unknowns"* by Walt Trybula and Deb Newberry, covers the facts that the vast majority of nanomaterial properties are unknown, and that their impact on people and the environment will not be known for some time. Chapter 7, *"What is Considered Reliable Information"* by Evelyn Hirt and Walt Trybula, addresses the fact that many sources of nano-safety information are not renewed on a regular basis and new regulations may change the existing recommendations. Consequently, there is a need to develop an understanding of reliable information sources. Chapter 8, *"Ethics and Communication: the Essence of Human Behavior"* by Craig Hanks, delves into the need for an ethical approach to nanotechnology safety, without which many questionable decisions might be made. Chapter 9, *"Behavior-Based Worker Safety for Engineered Nanomaterials"* by Christie M Sayes, James Y. Liu, and Matthew Gibb, addresses the issue that worker training and the implementation of controls for the application and handling of nanomaterials need to be thorough and built on the need for safety. Chapter 10, *"The Future of Nanotechnology Safety"* by Dominick Fazarro, provides a hypothetical view of the future, where safety is properly incorporated into the world of nanotechnology, and highlights the possible mechanisms for ensuring safety.

This author and the editors thank all the people who have been involved in this effort and is pleased to see the publication of this book, which represents years of development as nanotechnology has grown. Our goal with this book is to provide a basis for additional learning as the field develops and matures. No one organization or source has all the answers, but we hope that we have provided the reader with a starting point to ensure that he or she is creating and working in an environment that can be considered safe.

30 June 2023   Walt Trybula

Contents

Jitendra S. Tate and Roger A. Hernandez

Christie M. Sayes, James Y. Liu, and Matthew Gibb

Christie M. Sayes, James Y. Liu, and Matthew Gibb

Dominick Fazarro

About the Editors

Dominick Fazarro, PhD is an ATMAE Senior Fellow and professor in the Department of Technology at the University of Texas at Tyler. Dr Fazarro has been researching nanotechnology safety for over fifteen years. He was a co-principal investigator on federally funded programs with the National Science Foundation (NSF) and Susan Harwood to address nanomaterial safety education and training. Dr Fazarro is recognized as a senior member by the IEEE Nanotechnology Council. He is a member of the U.S. National Committee (USNC) Technical Advisory Group-IEC TC 113 (Nano-Electrotechnologies). He has published over 40 peer-reviewed articles, book chapters, and presentations on the topic of nanotechnology safety education. Dr Fazarro facilitated the development of an online nano-safety certificate program with a short course. In 2016, Dr Fazarro along with the grant team was recognized by the National Academy of Engineering Exemplars of Engineering Ethics Education.

Dr Christie M. Sayes is a practicing research scientist, consultant, and academician in the fields of toxicology, chemistry, material science, and environmental health. Currently, she holds the position of Associate Professor of Environmental Science & Toxicology at Baylor University (Waco, Texas). Sayes is a subject matter expert in the chemistry of materials, exposure science, health effects, and risk. Her activities include working with partners, collaborators, and trainees in designing studies related to safety-by-design considerations of engineered substances and emerging contaminants used in pharmaceutical, agricultural, and consumer products. Sayes is also interested in *occupational safety* and *environmental transformations* of chemical and particle systems in complex matrices.

Walt Trybula, PhD, MBA, IEEE Fellow, SPIE Fellow, IEEE Distinguished Lecturer, is a Director of the Trybula Foundation, Inc., co-founder of GeoSingularity, Inc., an Adjunct Professor in the Ingram School of Engineering at Texas State University, past Director of the Nanomaterials Applications Center at Texas State, and past Director of the Center for Emerging Technology Commercialization at the University of Texas at Austin. As a Senior Fellow of the Technical Staff at SEMATCH, he was involved in the development of semiconductor lithography as it moved into the nano-realm. Walt has been involved in various aspects of nanotechnology since 1979 and was an early driving force in Nano-Safety.

Dr Jitendra S. Tate, professor of manufacturing engineering at Texas State University, has established safe handling practices for industrial (such as nanoclay) and engineered (such as carbon nanotubes) nanoparticles in his research and teaching, dealing with advanced polymer nanocomposites. His research areas include developing, manufacturing, and characterizing the high-performance polymeric thermoplastics and thermoset nanocomposites for Thermal Protection Systems (TPS), rocket ablatives, fire-retardant interior structures of mass transit and aircraft, lighter and damage-tolerant wind turbine blades, fiber-reinforced high-temperature composites, replacement of traditional composites using bio-based materials, sustainable composites from renewable resources, cellulose nanofibers, conductive/magnetic/high-temperature polymers for 3D printing, nanotechnology education, and nanotechnology safety.

Dr Craig Hanks is NEH Distinguished Teaching Professor of the Humanities and professor of philosophy at Texas State University. A winner of seven teaching awards, including the highest recognition at Texas State University and the University of Alabama in Huntsville, he has more than 20 years of experience teaching philosophy of technology, engineering ethics, and professional ethics. He has served Texas State on the

https://doi.org/10.1515/9783110781830-203

Institutional Animal Care and Use Committee (IACUC) and the Institutional Review Board (IRB), which he chaired for three years. Dr Hanks was a visiting associate professor at the Stevens Institute of Technology and has offered short courses and seminars on philosophy and ethics for teachers, from high school through to doctoral programs. He is an active member of the International Society for Philosophy and Technology, a member of the editorial board for Philosophy in the Contemporary World, and an editor for the book series Philosophy of Engineering and Technology (Springer).

List of Contributing Authors

Eylem Asmatulu
Department of Mechanical Engineering
Wichita State University
1845 Fairmount St.
Wichita, KS 67260, USA
E-mail: e.asmatulu@wichita.edu
Chapter 2

R. Asmatulu
Department of Mechanical Engineering
Wichita State University
1845 Fairmount
Wichita, KS 67260-0133, USA
E-mail: ramazan.asmatulu@wichita.edu
Chapter 3

Dominick Fazarro
The University of Texas at Tyler
3900 University Blvd
Tyler, TX 75799, USA
E-mail: dfazarro@uttyler.edu
Chapter 10

Barbara Foster
7101 Royal Glen Trail
McKinney, TX 75072, USA
E-mail: bfoster@the-mip.com
Chapter 1

Matthew Gibb
Department of Environmental Science
Baylor University
#97266 One Bear Place
Waco, TX 76798-7266, USA
Chapter 5 and 9

Emily Kay Hanks
Department of Political Science
Texas State University
601 University Dr.
San Marcos, TX 78666, USA
Chapter 8

J. Craig Hanks
Department of Philosophy
Texas State University
601 University Dr.
San Marcos, TX 78666, USA
E-mail: craig.hanks@txstate.edu
Chapter 8

Roger A. Hernandez
Texas State University
601 University Dr.
San Marcos, TX 78666, USA
Chapter 4

Evelyn H. Hirt
Laboratory Planning & Performance
Management Directorate
Pacific Northwest National Laboratory
MSIN: K5-10
Richland, WA 99352-0999, USA
Chapter 7

W. S. Khan
NED University of Engineering and Technology
Faculty of Engineering
800MyHCT, 80069428 Fujairah, UAE
E-mail: wkhan1@hct.ac.ae
Chapter 3

James Y. Liu
Department of Environmental Science
Baylor University
#97266 One Bear Place
Waco, TX 76798-7266, USA
Chapter 5 and 9

Deb Newberry
Newberry Technical Associates
2984 S Herman St.
Milwaukee, Wisconsin 53207, USA
E-mail: deb.newberry@dctc.edu
Chapter 6

Christie M. Sayes
Department of Environmental Science
Baylor University
#97266 One Bear Place
Waco, TX 76798-7266, USA
E-mail: christie_sayes@baylor.edu
Chapter 5 and 9

Jitendra S. Tate
Ingram School of Engineering
Texas State University
601 University Drive
San Marcos, TX 78666, USA
E-mail: JT31@txstate.edu
Chapter 4

Walt Trybula
The Trybula Foundation
4621A Pinehurst Dr. South
Austin, TX 78747, USA
E-mail: w.trybula@tryb.org
Chapter 6 and 7

Barbara Foster
1 The World of Nanotechnology

Foreword

Although this chapter is being edited seven short years after its original publication, Nanotechnology and nanosafety have exploded. Not only has Nanotechnology given rise to intriguing new commercialized products, it has become a significant partner to the exponential growth in Artificial Intelligence (AI), the origin for new sciences, such as nanoinformatics, and a driving force behind global efforts, such as sustainability. The tragic world-wide epidemic of Covid-19, which began early in 2020, fostered a new push for Nanotechnology in medicine and biosensors. Many original questions, like "Is Nanotechnology a discipline unto itself or will it be subsumed into other sciences?" still remain. New ones, like "When will graphene really become commercial?" and "What is the role of Nanotechnology in the global movement toward sustainability?" have emerged. In light of these many changes, this updated chapter shifts its emphasis away from some of the original topics, such as a detailed review of the market, and focuses more on the impacts and results of these new drivers and enquiries.

1.1 Introduction

The year was 1944. A 23-year-old chemist stood at a laboratory bench, in a top-secret facility hidden in the New Mexican desert. His faced several challenges: First was the chemistry and metallurgy of refining the first nonmicroscopic quantities of plutonium for the war-crucial Manhattan Project. Second was keeping his laboratory safe from the unknown hazards of conducting research in the newly discovered realm of radioactivity.

A year later, he produced what he later called his "other major Manhattan Project"—me. Under my father's tutelage, I grew up to become a high school chemistry teacher My father's early research and lab practices exerted considerable influence on me: laboratory safety was always job #1. For example, I became famous for unexpectedly standing in the front of the class during laboratory sessions and loudly counting, "1, 2, 3 . . . ." My students knew I was counting faces, looking carefully to see who was NOT wearing their safety goggles. Also, during those years, the promise of broad peace-time use of the atom intrigued me and, as my career developed, both the impact and risks of the Atomic Age found their way into my curricula.

Acknowledgement: I would like to thank Chase Malone and Joshua Berger, University of Texas at Tyler graduate students for their assistance in researching chapter 1.

https://doi.org/10.1515/9783110781830-001

For the past 40 years, I have been a technical marketing manager and strategic consultant to the microscopy arena, helping companies identify and launch new technologies, many of which are used to image, characterize, and measure the nanoworld.

Why are these two historical notes important? First, as a "daughter of the Atomic Age," I see many similarities between the promise, evolution, and risks of nuclear science and those of Nanotechnology. Second, my in-depth involvement in microscopy and "nanoscopy" literally gave me a unique view into the nanoworld. In that role, I have had an opportunity to work with many of the cutting-edge techniques in atomic force, confocal, super-resolution, and even light microscopy, allowing me to actively observe the nanoworld at work. From those perspectives, I share with you this overview of Nanotechnology.

How does Nanotechnology compare to those early days of atomic research? As in the early days of working with radioactive materials, there is much we do not know about the interaction between nanomaterials and the biome in general, and humans in particular. Walt Trybula and Deb Newbury discuss this very issue in Chapter 6 of this book, "Understanding the implications of nanomaterial unknowns."

Early pioneers in radioactivity had dosimeters and Geiger counters, but what can be used to detect exposure to nanomaterials? Risk is measured by both exposure to the hazard and the dose. Although radiation is relatively easy to detect and measure, the new throng of nanoparticles makes developing selective nanosensors challenging.

Not only are the large numbers and variety of nanomaterials daunting, to add even further complexity, sometimes a small difference in structure produces a major shift in potential safety. Carbon nanotubes, described in greater detail in Section 1.3.1, are a good case in point. Observations suggest that if they have closed ends, they tend to orient themselves parallel to surfaces such as skin and are benign. However, if they have open ends, they tend to approach surfaces head on, potentially penetrating skin and causing havoc inside cells.

A second interesting comparison between radioactive materials and nanomaterials is the rate and extent of commercialization. In both research and commercialized products, there is potential for exposure during synthesis, modification, processing, characterization, and testing. Even now, 80-odd years after the Manhattan Project, radioactive materials remain largely in the purview of either the government or developers of nuclear power plants and nuclear medicine. In comparison, Nanotechnology has moved aggressively into the mainstream. Within the last decade, it has undergone explosive commercialization across a broad landscape from energy to agriculture, golf clubs to computers, and medicine to food processing and water treatment. At the original of writing this chapter, The Nanotech Industries Association [1] reported over 2000 member companies, and the Nanotech Database of Products [2] cited over 2300 entries. Today, Nanotech Database lists nearly 5500 products and a sister site, StatNano, nearly 11,000!

As in the mid-1940s, we stand on the brink of a brave new world, with nanomaterials promising blazing leaps in technology. However, there is a great deal for us to learn about how best to use them safely and ethically. Contributions to this book were invited specifically to help you begin that investigation. In this chapter, we begin with the fundamentals: the technology, its applications, current directions in research, the market, and a glimpse of the future.

1.2 What is Nanotechnology?

Long before Nanotechnology and the science to investigate it came into existence, man unknowingly put the power of tiny particles to work, using colloidal gold to make the magnificent stained glass windows and goblets of the Middle Ages and the unique Wootz steel in Damascene swords.

The nanoworld is the world of the very small and is typically defined by structures and unique interactions that occur at the size range of 1–100 nm. For a quick size comparison, a nanometer is about 100,000th the diameter of an average human hair.

The concept of technology at the nanoscale was first hypothesized in the abstract by Dr. Richard Feynman in his 1959 presentation to the annual meeting of the American Physical Society at the California Institute of Technology. In "There's Plenty of Room at the Bottom," [3] he began by posing the question, "How small can we write?" More specifically, he asked, "Could we write the entire Encyclopedia Britannica on the head of a pin?" After explaining that we can actually write much smaller, he moved on to "How can we *read* what we have written?" It was the first conceptual step toward manipulating and controlling things at the nanoscale. The use of photolithography to print semiconductors was in its infancy [4] and, although the electron microscope had been invented, it would be decades before we had the high-resolution electron microscopy and, eventually, atomic force microscopy needed to really "read" the nanoworld.

In 1972, Taniguchi coined the term "nanotechnology" to describe semiconductor processes, such as thin film deposition and ion beam milling, which could exert control at the nanoscale. His definition was as follows: "'Nanotechnology' mainly consists of the processing of, separation, consolidation, and deformation of materials by one atom or one molecule." However, the term did not catch on until the 1980s. That period saw a convergence of thought and technology. Eric Drexler, who was unaware of Taniguchi's prior use of the term, published his first paper on Nanotechnology in 1981, followed in 1986 by his foundational work, *Engines of Creation* [5]. In 1984, Feynman updated his earlier presentation in "Tiny Machines." [6]. His slides show just how much the semiconductor industry has changed.

In 1981, scanning tunneling microscopy (STM) was invented, giving the world not only a unique view into the nanoworld but also enabling manipulation and measurement at that scale. In 1986, fullerenes were discovered, and nanotech reached a tipping point. The race was on!

1.3 The Growing World of Nanomaterials

Interesting things happen to chemistry, physics, and biology at the nanoscale. Surface area increases millions of times. There are changes in hydrophobicity and hydrophilicity, strength, transparency, melting point, and the ability to conduct both heat and electricity. Gold is no longer gold, but red. Perhaps most amazing of all, materials have the ability to self-assemble. Chapter 2, "The world of engineering nanomaterials" by Eylem Asmatulu, discusses this topic further, but here is a quick overview.

1.3.1 Carbon-Based Nanomaterials

Graphene [7] has sparked considerable interest. It is a two-dimensional (2-D) material comprised of single sheets of hexagonally bonded carbon only one atom thick. In Figure 1.1, it would be just one, single layer of the material graphite. It is flexible and transparent, conducts electricity better than silver, conducts heat better than diamonds, and is stronger than steel. Ironically, this exotic material was discovered using rudimentary tools. In 2004, Andre Geim and Konstantin Novoselov (both at the University of Manchester, UK) used scotch tape to pull layers of graphene off a piece of graphite (essentially, pencil lead) until they achieved a layer one atom thick—the first 2D material. Interestingly, as pointed out by Adrian Nixon [8], editor of the Nixene Journal, prior to 2004 the existence of 2D materials was considered "impossible."

The efforts of these two researchers were quickly rewarded. Only six short years later, they received the Nobel Prize in Physics [9]. During those early years, growing even a single crystal of graphene for research purposes was a challenge. Today, there are over 40 companies who produce graphene on a regular basis.

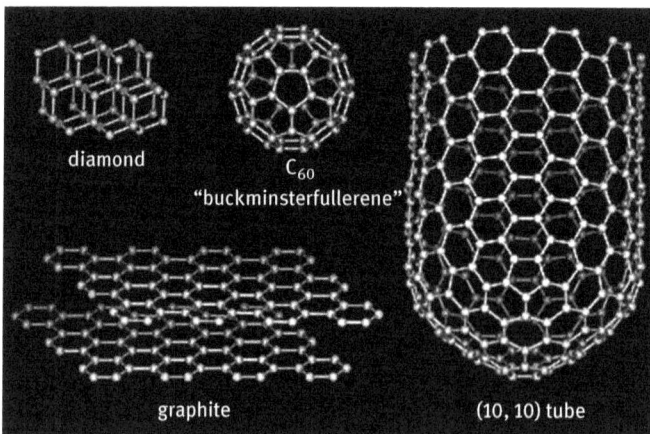

Figure 1.1: Relationships between various allotropes of carbon [7].

Structurally, graphene forms the building blocks for a number of other structures. Although not manufactured in this fashion, when a sheet of graphene is shaped into a ball (Figure 1.1) it generates a fullerene (bucky ball, C_{60}). Rolled into a tube, it produces a single-walled carbon nanotube (SWCNT, also Figure 1.1), with wall thickness of the order of 1–2 nm. Covering a SWCNT with one or more additional sheets of graphene produces double- or multi-walled CNTs (DWCNTs or MWCNTs, respectively). CNTs can also aggregate, forming nanohorns and nanobrushes [10], structures that are projected to have importance in new batteries and capacitors.

One of the latest products is wrinkled graphene. Recent research describes how bacteria are tucked under a blanket of graphene [11]. After heat and vacuum are applied, the graphene essentially shrink-wraps the bacteria, generating what the inventors term "a new carbon allotrope: a half CNT linked to graphene." The resulting wrinkles are about 7 nm in height and exhibit anisotropic electrical properties along the length of the wrinkle versus across the wrinkle.

The question as to who actually discovered CNTs is a matter of hot debate that has intellectual property implications [12–14]. Nisha and Mahajan [15] point out that Mother Nature has been producing CNTs for eons, and they can be found both in volcanic byproducts here on earth and in materials recovered from space.

Although carbon filaments and hollow fibers have been observed in the laboratory for over a century (Figure 1.2), it is widely reported that the first definitive images of CNTs were electron micrographs of MWCNTs, published in 1952 by Radushkevich and Lukyanovich in a Russian publication [16]. However, because it was published in Russian, it was not widely read. CNT research increased through the 1970s and 1980s, some of which resulted in patents. However, the concept did not gain much traction. Nisha and Mahajan propose a number of reasons for the lack of interest, including the fact that the CNTs were structurally imperfect and, as a result, displayed no interesting properties of commercial value. Further, they cite that early studies focused more on growth mechanisms rather than on discovering new carbon structures, and that researchers were limited by the then-available analytical tools.

In 1991, that scenario changed. Sumio Iijima of NEC developed a high-resolution tunneling electron microscopy and began actively researching a wide variety of carbon materials, which resulted in publication of a seminal paper on MWCNTs in *Nature* [17]. He later went on to image and describe SWCNTS. Because of the breadth and depth of his work, Iijima's research became the cornerstone of carbon nanostructure research and, as a result, he is often cited as the inventor of CNTs [17].

1.3.2 Colloidal-Based Nanomaterials

A second family of nanomaterials can be found in the colloidal world. Colloidal silver has been known since the mid-1880s but is now enjoying vast new markets in Nanotechnology. There are now nanosilver wipes, sponges, food storage materials, etc. Both nanosil-

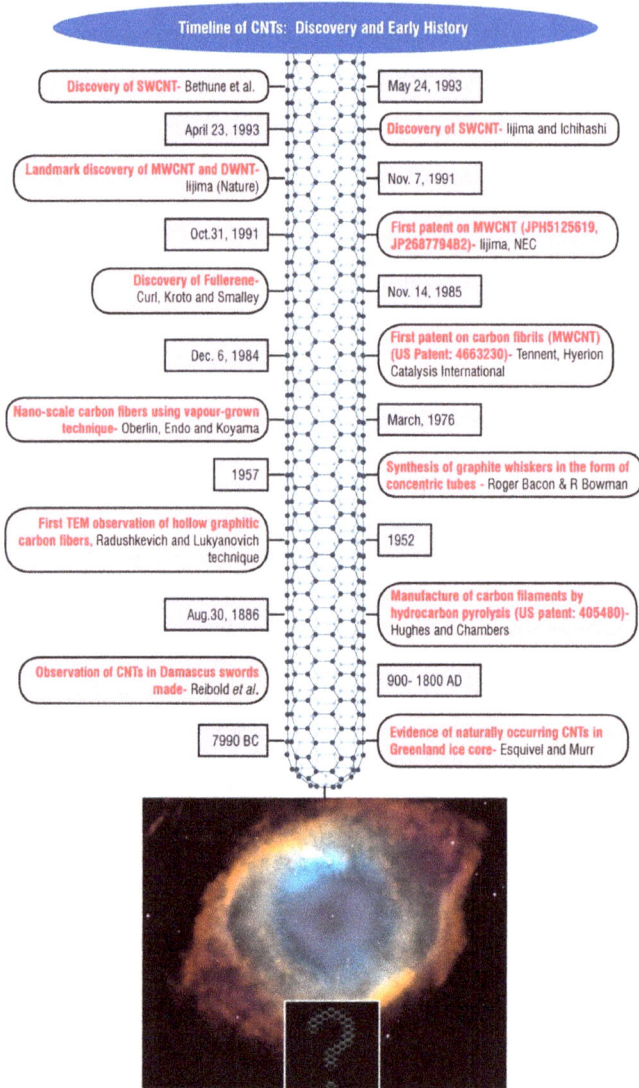

Figure 1.2: The history of carbon nanotubes (Courtesy, Nisha and Mahajan).

ver and nanogold are finding key applications in Raman spectroscopy on both substrates (surface-enhanced Raman spectroscopy, SERS) and atomic force microscopy probe tips (tip-enhance Raman spectroscopy, TERS).

1.3.3 Quantum Dots

Quantum dots [18] (QDs) form a third group of nanomaterials. These optoelectronic nanoparticles can be made of a variety of semiconductor materials, including silicon, cadmium selenide, cadmium sulfide, and indium arsenide. Their properties change as a function of both size and shape. As discussed in *Wikipedia* [19], "larger QDs (radius of 5–6 nm, for example) emit longer wavelengths resulting in emission colors, such as orange or red. Smaller QDs (radius of 2–3 nm, for example) emit shorter wavelengths resulting in colors such as blue and green, although the specific colors and sizes vary depending on the exact composition of the QD."

This article goes on to describe their range of applications: "Because of their highly tunable properties, QDs are of wide interest. Potential applications include transistors, solar cells, LEDs, diode lasers and second-harmonic generation, quantum computing, and medical imaging [20]. Additionally, their small size allows QDs to be suspended in solution, which leads to possible uses in inkjet printing and spin-coating. These processing techniques result in less expensive and less time-consuming methods of semiconductor fabrication." [20]

Because of their excellent photoelectric efficiency, QDs are being proposed for the next generation of solar cells. For example, normal solar cells convert each photon into an electron. In comparison, QDs can convert each photon into two electrons. In microscopy, their tunability and stability makes them an interesting substitute for more sensitive and degradable organic fluorophores. The same fluorescence makes them interesting candidates for electronic displays. For example, different sized QDs can be grouped to make the R-G-B dots that form a pixel.

1.3.4 Biologically Based Nanomaterials

As illustrated in Figure 1.3 below, there is a growing family of biological nanoparticles [21]. Although this particular image refers to immunoregulation, similar nanoparticles are being actively researched and developed to locate, diagnose, treat, and monitor many diseases. As discussed in the analysis of the market (Section 1.8), drug delivery ranks high in biological nanoparticle applications.

1.4 Instrumentation for Investigating Nanotechnology

Although other chapters in this book are more specific about methodology, a quick review of some of the techniques for imaging nanomaterials is useful here.

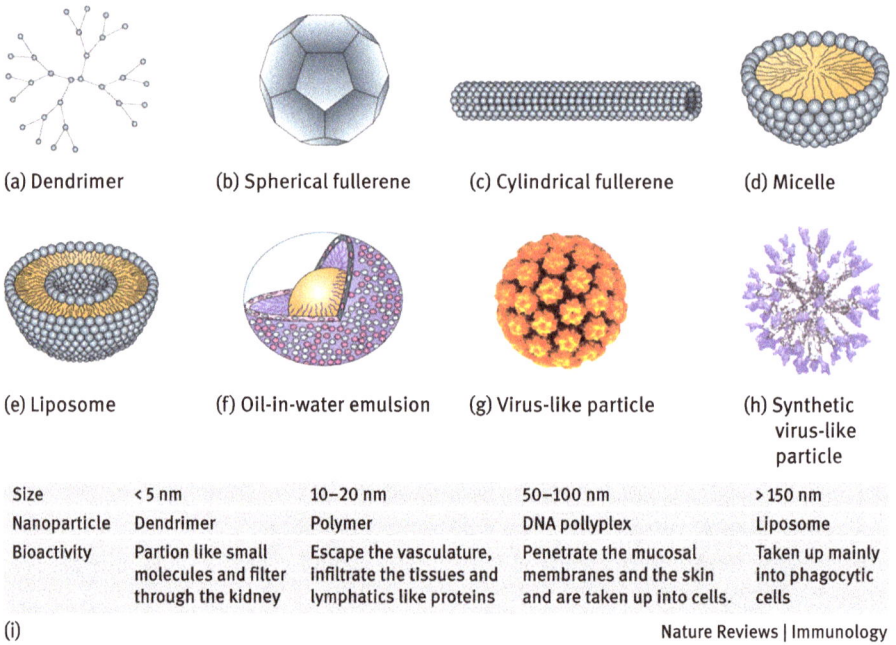

| (a) Dendrimer | (b) Spherical fullerene | (c) Cylindrical fullerene | (d) Micelle |

| (e) Liposome | (f) Oil-in-water emulsion | (g) Virus-like particle | (h) Synthetic virus-like particle |

Size	< 5 nm	10–20 nm	50–100 nm	> 150 nm
Nanoparticle	Dendrimer	Polymer	DNA pollyplex	Liposome
Bioactivity	Partion like small molecules and filter through the kidney	Escape the vasculature, infiltrate the tissues and lymphatics like proteins	Penetrate the mucosal membranes and the skin and are taken up into cells.	Taken up mainly into phagocytic cells

(i)

Nature Reviews | Immunology

Figure 1.3: Nanotechnology topologies that can be applied to immunoregulation include nanoparticles (a–c), nanoemulsions (d–f) and virus-like particles (g–h). (Nature Reviews | Immunology).

From the early days of Nanotechnology, transmission electron microscopy (TEM) [22] has been a key technique for investigation, well suited because of its ability to image at the atomic and molecular level. TEM transmits an electron beam through the material of interest and is able to magnify 1000 to 1,000,000 times. The beam is preferentially absorbed by both the density of the material and its thickness, resulting in shifts in both amplitude (intensity) and phase that generate contrast and provide information about particle morphology and surface, thickness of surface coatings, and even atomic order.

Scanning electron microscopy (SEM), as the name implies, uses an electron beam to scan the surface. By using a variety of detectors and a range of electron voltages, images are created using either backscattered electrons or secondary electrons, revealing phases within the material and also the morphology and texture of the surface. SEM devices are often fitted with energy dispersive X-ray (EDX) detectors for elemental analysis and with diffracted backscattered electron detectors (EBSD) to examine crystallographic orientation. SEM can typically obtain magnifications ranging from several thousand to 30,000×. Desktop models that integrate optical microscopy with electron microscopy seamlessly bridge conventional optical magnifications of 20–1000×through SEM ranges of up to 20,000×. Furthermore, these instruments can combine brightfield or fluorescence in the optical range with SEM and EDS in the electron microscopy range.

Scanning probe microscopy (SPM) includes two families of technology: (1) STM and nearfield scanning optical microscopy (NSOM), which take advantage of nearfield effects, and (2) atomic force microscopy (AFM), which scans the surface with an atomically sharp tip. Depending on the application, the scans can be done with the tip in contact with the surface, hovering over the surface, or "hopping" or "tapping" across the surface. SPM offers over 40 different imaging modes, with contrast generated by changes in topography, phase, hardness, tacticity, various electrical parameters (voltage, capacitance, tunneling current), magnetism, etc. In addition to "images," these modes can provide quantum mechanical testing.

The latest iteration of these microscopes combines AFM with Raman spectroscopy. Raman has always been a challenge because of its weak signal. Raman at the nanoscale presents even more of a challenge. However, the signal can be boosted nearly a million-fold (enhanced Raman spectroscopy) by using materials such as gold and silver to enhance the electric field at the tip–sample interface. In conventional Raman, this goal is achieved by using special substrates to generate surface-enhanced Raman spectroscopy (SERS). In AFM, special tips coated with nanogold, or similar materials are used, generating tip-enhanced Raman spectroscopy (TERS). Until recently, TERS was difficult to control experimentally. However, new hybrid instruments [23] have stabilized the AFM–spectrometer interface, facilitating the use of TERS in both research and more routine applications. Figure 1.4 shows a true AFM/Raman hyperspectral image, in which both the AFM image and the Raman signal were collected simultaneously from a single CNT. The ability to combine both molecular-level imaging with the power of chemical analysis in an easy-to-use system has powerful implications for nano-safety.

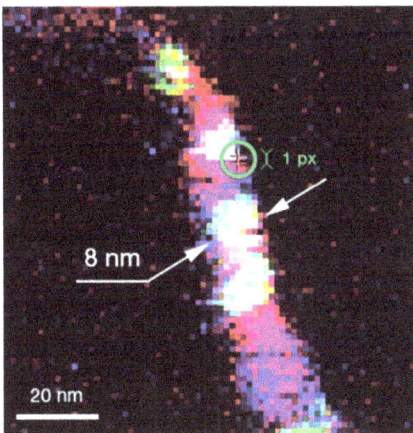

Figure 1.4: Hyperspectral nanoscale Raman image of an isolated carbon nanotube. Raman resolution 8 nm; scan area 100 nm × 100 nm; scanned in 1.2 nm steps at 100 ms/pixel; scan time < 9 min. (Curtesy of HORIBA and AIST-NT).

When it comes to the nanoworld, optical microscopy is not typically included in the analytical scheme. With a limit of resolution of about 250 nm, it does not seem to be a good fit; however, several approaches can be applied. Scanning white light interferometry (SWLI) and phase shifting interferometry (PSI) offer the ability to image nanoscale variations in step height and topography. Surface-enhanced ellipsometric contrast microscopy (SEEC) is a more recent entrant into the field, combining reflected light differential interference contrast with the use of special substrates. SEEC routinely images step heights of the order of a few nanometers.

Although imaging nanosteps in the Z plane has been available for over 40 years, using optical technology to break the 100 nm barrier in the XY plane has been more of a challenge. The past decade has seen the rise of a variety of super-resolution techniques, used primarily in the life sciences [24]. Some, such as structured illumination microscopy (SIM), stimulated emission depletion (STED), and *light-sheet microscopy* manipulate illumination to drop the conventional XY resolution to the 65–90 nm range.

Others, like photoactivation localization microscopy (PALM) and stochastic optical reconstruction microscopy (STORM), both taking advantage of new photo switchable probes, locate and map individual molecules. From those maps, it is possible to "image" structures and processes at resolutions on the order of approximately 20 nm. These instruments, however, typically fall in the $400,000–$800,000 range.

Another, elegantly simple solution comes from CytoViva and has applications in both life sciences and materials sciences. The device consists of modified light source and condenser system that readily retrofits to most current microscopes. It makes use of near-field effects to drop XY resolution to about 90 nm and detection below 50 nm (Figure 1.5) for under $20,000. It can also expand into a larger, hyperspectral imaging system (~$125,000).

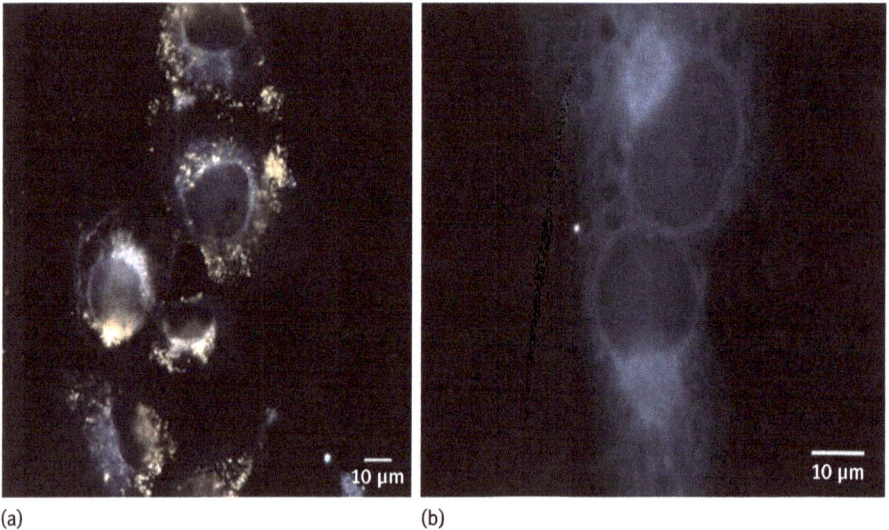

(a) (b)

Figure 1.5: Gold nanoparticles used as indicators in cancer cells: (a) Gold nanoparticles taken up by cancer cells. (b) Negative/control. (Courtesy of CytoViva).

1.5 Where is Nanotechnology Today?

Today, Nanotechnology is hitting its stride, with active education, research, and development on myriad fronts that are already making dramatic changes in how we live, what we wear, the food we eat, our healthcare, and how we manage resources such as energy and water.

For over a decade, industry observers have asked the question, "Can Nanotechnology continue to exist as a separate entity, or will it devolve back into the basic sciences of chemistry, physics, and biology/medicine?" As discussed below in Section 1.6 on applications and 1.8 on the Nanotechnology market, there is ample evidence to suggest that Nanotechnology has firmly established its own identity, both scientifically and commercially. That said, it also has deep tentacles in primary sciences including chemistry, physics, biology, and medicine.

The COVID pandemic was good news/bad news for Nanotechnology. Bad in the sense that, as discussed below (Section 1.7: Role of the government), funding has been slowed. Good in the sense that it played a pivotal role in Operation Warp Speed's ability to quickly design, manufacture, and deliver a vaccine that was effective in limiting the severity of the virus and opening a door to rapid development of other drugs and therapies based on nano delivery.

Today's Nanotechnology is cutting edge and highly multidisciplinary, reaching across the scientific and informational spectrum from biology and medicine to materials science and semiconductors, to practical applications in energy generation/ transmission/storage, agriculture, road surfaces and water purification. It has also spawned interesting offspring, including the new science of nanoinformatics and an intriguing new direction in materials development called Meta Materials [25], in which nanomaterials are combined or layered with other nanomaterials, semiconductors and/or metals to make materials unlike any we have seen before. Mxenes are one such material (Figure 1.6). They prompted our editor, Walt Trybula, to ponder in the February 2023 issue of his Nano-blog.com: "More nanomaterials, or is it metamaterials, or semiconductors?" [26]. Here is the schematic of a Mxene structure from that blog.

Nanotechnology also cuts across many platforms. Consider, for example, the interplay between private enterprise, government, academia, and foundations across the globe. As will be discussed in the next sections, in the US, the National Nanotechnology Initiative (NNI) is driving strong development and cooperation across academe-government-industry efforts. Globally, it is a key enabler for the UNs new Sustainable Development Goals (SDGs). In these efforts, Nanotechnology is seen as "converging": unifying and bringing multiple technologies together to enable greater productivity and advancement.

The bottom line is that a student or practitioner can enter Nanotechnology through many gateways. Nanotechnology only requires that you bring to a sense of cooperation, team building, and the ability to think outside the box. Just choose a discipline, choose

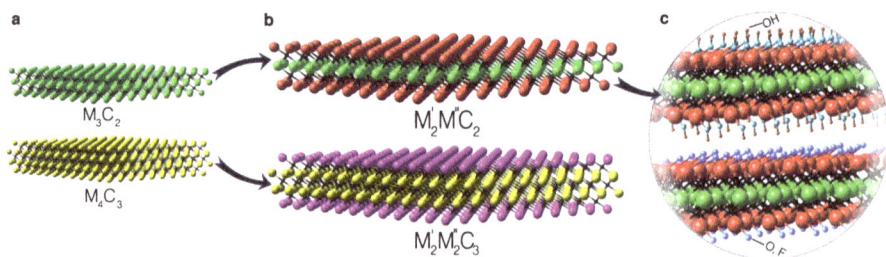

Figure 1.6: "Schematics of the new MXene structures. (a) Currently available MXenes, where M can be Ti, V, Nb, Ta, forming either monatomic M layers or intermixing between two different M elements to make solid solutions. (b) Discovering the new families of double transition metals MXenes, with two structures as M′2M″C2 and M′2M″2C3, adds more than 20 new MXene carbides, in which the surface M′ atoms can be different from the inner M″ atoms. M′ and M″ atoms can be Ti, V, Nb, Ta, Cr, Mo. (c) Each MXene can have at least three different surface termination groups (OH, O, and F), adding to the variety of the newly discovered MXenes" [27].

whether you would prefer research and/or commercialization, and Nanotechnology will provide a pathway into the future. The next section details some of the more active application areas.

1.6 Applications

StatNano.com is an excellent place to get oriented about applications in Nanotechnology. Their "Products" section (outlined in Table 1.1) reveals unexpected nooks and crannies far beyond those normally discussed. For example, while Nanotechnology is widely associated with electronics, energy, textiles, and sports equipment, it is less often connected with cosmetics, construction, food, and home appliances.

Table 1.2 details some of the subsets from the previous list.

StatNano also provides a geographic distribution of 68 countries in which these products are produced. Figure 1.7 summarizes data from Table 1.1 across the top 10 producer countries. This data was updated in March 2022. It shows that the US is the most prolific country, globally, in the array of nanoproducts it produces, followed by China and Germany. Whereas high tech centers, such as Japan, South Korea, and India were expected, Iran was a surprise entrant in this Top 10 List.

Figure 1.8 illustrates how key nanomaterials, ranging from nano-silver to graphene and tungsten sulfide are used across those same application areas.

One of the most important product categories is the family of Nanoparticles. StatNano categorizes them as either 1D, 2D, or 3D "nano-objects" or as "nano-structures. Here is their description of how the Nanomaterials Database was founded and is organized:

"Nanotechnology is based on the manipulation and control of nanomaterials, and in fact nanomaterials are the cornerstones of the new technology. "*Nanomaterials*

Table 1.1: The products listing at StatNano.com reveals a wide array of nano applications. In addition, data are broken down by the number of products in the listing, the number of companies that make that type of product, and the number of countries in which those products are manufactured. (Data compiled by the author from multiple listings at (https://product.statnano.com/)).

Products	# Products	# Companies	# Countries
Electronics	1957	157	30
Construction	1154	612	49
Medicine	1296	562	48
Cosmetics	999	330	32
Textiles	886	555	42
Automotive	825	315	41
Renewable Energy	438	273	36
Petroleum	317	142	32
Environment	607	290	35
Food	423	195	32
Home Appliance	379	169	29
Agriculture	244	88	28
Sports & Fitness	164	57	5
Printing	203	111	22
Others	961	545	44

Table 1.2: Further detail of Nanoapplications from StatNano.com.

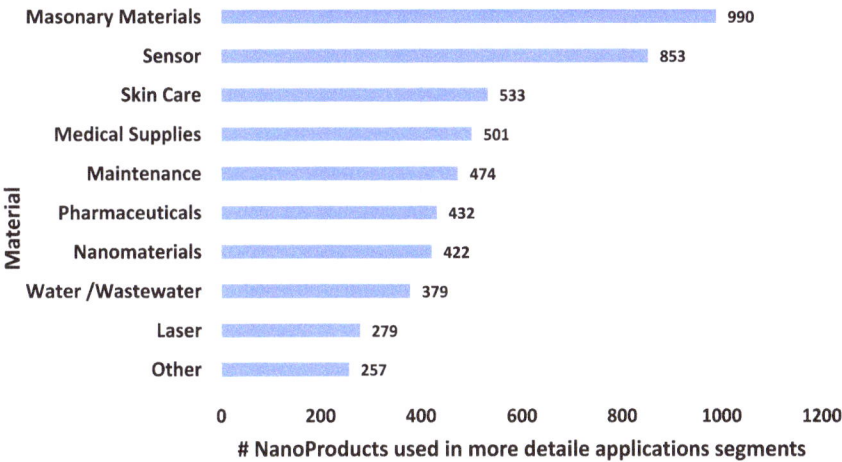

Material	# NanoProducts
Masonary Materials	990
Sensor	853
Skin Care	533
Medical Supplies	501
Maintenance	474
Pharmaceuticals	432
Nanomaterials	422
Water /Wastewater	379
Laser	279
Other	257

NanoProducts used in more detaile applications segments

Database" was launched in 2016 with the aim of introducing different types and morphologies of nanomaterials and monitoring the publications, patents, and commercialization trend of these materials. Mission of the database is collecting, analyzing, and publishing the information on nanomaterials in various aspects of science, technology, and commerce."

Electronics ○ Medicine ○ Construction ○ Cosmetics ○ Others ○ Textile ○ Automotive ○ Environment ○ Renewable Energies ○ Food   Home Appliance   Petroleum

Agriculture   Printing   Sports and Fitness

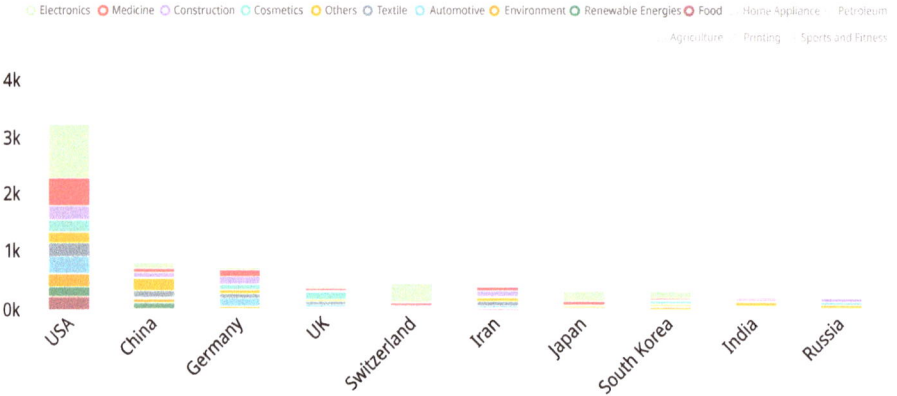

Figure 1.7: Listing of key countries producing the product categories outlined in Table 1.3.

Silver ○ Silicon dioxide ○ Titanium dioxide ○ Graphene ○ Tungsten disulfide ○ others

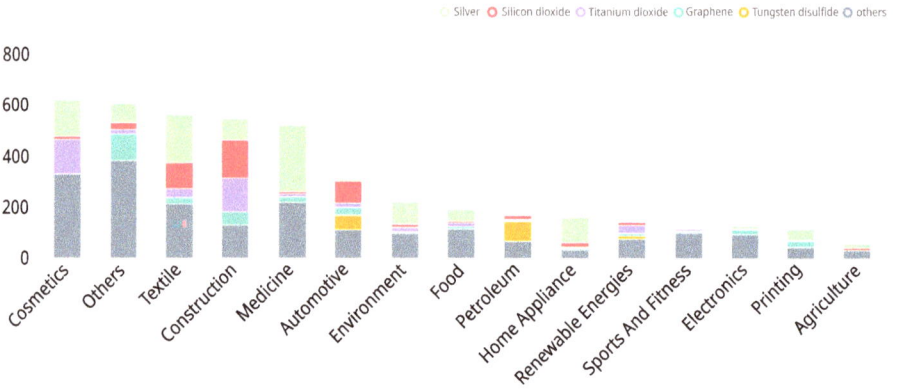

Figure 1.8: Cross tabulation of the major product categories outlined in Table 1.3 versus the key nanomaterials used.

This database is massive. It contains dynamic links to 54 different morphologies. It also refers to their studies of the Web of Science, indicating that these 54 morphologies account for 2,040,200 articles, 229,744 patents, and 4330 products. It is a well-organized "first step" for learning about each of these nanomaterials.

According to StatNano, the nanomaterials with the most views are graphene, borophene, and nanoparticles (Table 1.3) and those with the greatest growth rates in patents are MXenes, borophene, and nanofluids (Table 1.4) [28].

In summary, Nanotechnology is so universal in scope and wandering through the nanolandscape is intriguing. Each reader will have his/her own interests, so, in addition to StatNano.com, here is a list of URLs where you can research the latest and greatest applications:

–   www.nano.gov: Site for the US National Nanotechnology Initiative (NNI). See especially, "Big things from a tiny world" [30] and "Nanotechnology and energy: powerful things from a tiny world" [31] at http://www.nano.gov/node/734.

Table 1.3: Most viewed nanomaterials at StatNano.

Graphene	Borophene	Nanoparticles
Articles: 165,689	Articles: 737	Products: 3085
Patents: 21,329	Patents: 114	
Products: 307	Products:	

Table 1.4: Nanomaterials earning largest number of new patents [29]. (Nanofluids image: J. Appl. Phys. 113, 011301 (2013); doi: 10.1063/1.4754271).

MXene	Borophene	NanoFluids
Articles: 5983	Articles: 737	Articles: 27,931
Patents: 139	Patents: 114	Patents: 266

- www.Nanowerk.com: Nanotechnology Spotlights, especially some of their articles on food, energy, and water purification.
- www.AzoNano.com: Online publication for the nanocommunity.
- www.ACSNano.org: The American Chemical Society's journal dedicated to Nanoscience and Nanotechnology.
- www.spie.org/newsroom/nanotechnology: SPIE's nanotechnology newsroom.
- www.rdmag.com/topics/nanotechnology: R&D magazine's section on topics in Nanotechnology.
- Nano-Blog.com (Walt Trybula, editor): valuable insights from a Nanotechnology futurist.

1.7 The Role of the Government in Promoting Nanotechnology

Governments from around the world are driving the development, expansion, and even commercialization of Nanotechnology. Obviously, their first role is funding.

1.7.1 US Funding

In the USA, the Nanotechnology efforts of 20 different agencies and departments are coordinated under the *National Nanotechnology Initiative (NNI)*, 11 agencies of which are directly involved in science, engineering, R&D, and technology and the rest having Nanotechnology-related mission interests or regulatory responsibilities. [32] Reflecting the current status of discipline, NNI is tasked with supporting both research and technology transfer. Initiated in 2001, it has budgeted nearly $24 billion for nanoscience and technology over its lifetime.

Table 1.5 shows the actual and proposed budgets for FY2021 through FY2023. Note the impact of COVID in the decreases in 2022 and 2023.

The President's FY2023 budget allocates $1.99 billion in NNI funding [33]. Table 1.5 describes the allocation by agency, whereas Figure 1.9 shows the allocation by program component area (PCA). According to the government news release, the current budget represents "continued investment in the foundational research that will fuel new discoveries and application-driven research and development to advance technologies of the future and address world challenges".

In 2023, just as this book is being revised, NNI issued the FY 2024, outlining continued expanded investment. This comparison (Table 1.6) is organized by some of the key *strategic* efforts, such as information technology (CISE) and international partnerships (OISE).

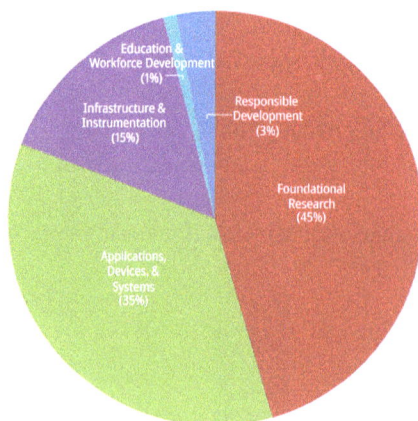

Figure 1.9: NNI Funding, FY 2023, allocated by "Program Component Area" (PCA).

Table 1.5: Allocation of the NNI budget from FY2021–2023, by agency (dollars in $million).

Agency	2021 Actual	2022 Estimated[*]	2023 Proposed
CPSC	0.4	0.4	1.2
DOC/NIST	62.7	62.2	61.0
DOD[**]	218.3	323.5	276.7
DOE[***]	377.7	385.0	409.7
DOI/BSEE	0.0	0.5	0.0
DOI/USBR	0.3	0.3	0.0
DOJ/NIJ	0.9	1.5	1.7
DOT/FHWA	0.2	0.2	0.5
EPA	2.5	4.1	2.7
HHS (total)	2457.2	1124.8	766.5
BARDA	1708.8	352.1	0.0
FDA	10.0	9.6	10.8
NIH	729.3	754.0	746.5
NIOSH	9.1	9.2	9.2
NASA	12.6	13.2	10.4
NSF	620.7	424.5	432.3
USDA (total)	23.4	22.6	26.2
ARS	4.0	4.0	4.0
FS	3.7	2.9	3.2
NIFA	15.7	15.7	19.0
TOTAL	3776.9	2362.7	1988.8

[*] 2022 numbers are based on appropriated levels.
[**] Funding levels for DOD include the combined budgets of the Air Force, Army, Navy, Defense Advanced Research Projects Agency, Defense Threat Reduction Agency, Office of the Under Secretary of Defense for Research and Engineering, and Joint Program Executive Office for Chemical, Biological, Radiological, and Nuclear Defense.
[***] Funding levels for DOE include the combined budgets of the Office of Science, Office of Energy Efficiency and Renewable Energy, Office of Nuclear Energy, and Office of Fossil Energy and Carbon Management.

1.7.2 Global Policy

Governments across the globe actively support nanoscience and technology. To illustrate how broadly, StatNano.com offers an extensive library of over 750 Policy Reports [35]. In this sampling, notice the wide range of geographies and topics.

- NASA Technology Roadmaps: TA 10: Nanotechnology (2015–2035)
- National Nanotechnology Policy & Strategy 2021–2030 Malaysia
- Nano4Society-NanoVision-2030 (2020–2030, The Netherlands)
- EU US Roadmap Nanoinformatics 2030 (2017–2030, Germany)
- Strategic Research and Innovation Agenda for Nanomedicine 2016–2030 (EU)
- Nano 2030: Manufactured Nanomaterials by 2030; Workplace Health and Safety Consequences in Small Businesses in France (2014–2030)

Table 1.6: NNI Nanotechnology Budget, FY2022–2024, broken down by key strategic efforts [34]. Note the contribution of the Disaster Relief Supplement and the CHIPS act to FY 2023, boosting the estimated base by nearly $5million. (BIO=(Directorate of) Biological Sciences; CISE = Computer and Information Sciences; EDU = Formerly known as Education and Human Resources or HER; MPS = Mathematical and Physical Sciences; SBE = Social, Behavioral and Economic Science; TIP = Technology, Innovation and Partnerships; OISE = Office of International Science & Engineering).

	National Nanotechnology Initiative Funding[1] (Dollars in Millions)					
	FY 2022 Actual[2]	FY 2023 Estimate Base	Disaster Relief Supplemental (CHIPS and		FY 2023 Estimate Total	FY 2024 Request
			Base	Science		
BIO	$39.95	$39.95	–	–	$39.95	$39.95
CISE	15.62	14.05	–	–	14.05	14.05
EDU	10.85	2.50	–		2.50	2.50
ENG	267.13	190.95	–		190.95	231.75
MPS	206.17	126.00	–	–	126.00	126.00
SBE	0.40	0.40	–	–	0.40	0.40
TIP	4.90	5.15	2.51	2.39	10.05	13.54
OISE	0.10	0.10	–	–	0.10	0.10
Total	$545.12	$379.10	$2.51	$2.39	$384.00	$428.29

[1] Funding displayed may have overlap with other topics and programs.
[2] FY 2022 Actuals may be greater than future fiscal years due to the receipt of more meritorious programs than expected.

– Towards an African Nanotechnology Future: Trends, Impacts and Opportunities 2020–2030, UN Economic Commission for Africa World
– EU 2030 Strategic Plan for Nanofabrication: a NanoFabNet Roadmap (2022–2030, EU)
– European Roadmap for Graphene Science and Technology (2013–2030, EU)
– Read more: https://statnano.com/policydocuments#ixzz86LQRJU43

Some of these reports are simply a review or evaluation of on-gong projects. Others detail strategies or even full plans & programs. Others discuss official acts. However, all of them offer projects well out into the future, often to 2030 or further. To make the library even more useful, StatNano has subcategories for National/International/Subnational/multinational. Also, they break the collection down by ELSA-SE, Safety, Health, Environment, Social, and Ethical.

1.7.3 Impact of the UN

In 2005, the UN convened a special Task Force on Science, Technology, and Innovation to address specifically the potential of Nanotechnology for sustainable development. Their findings were reviewed in an excellent *PlosMed* article, "Nanotechnology and the developing world." [36] Although this research was conducted over a decade ago, the results provide an important perspective on the extensive real-life world of Nanotechnology today.

The task force recruited 85 Nanotechnology experts from around the world. Of those, 63 completed the three-round process, with 60 % of that group representing developing countries, and the other 40 % representing developed countries. The question posed to them was: "Which do you think are the nanotechnologies most likely to benefit developing countries in the areas of water, agriculture, nutrition, health, energy, and the environment in the next 10 years?" They were asked, specifically, to evaluate their answers in terms of the following factors, which are still highly relevant in terms of assessing the potential for Nanotechnology:

– *Impact*: How much difference will the technology make in improving water, agriculture, nutrition, health, energy, and the environment in developing countries?
– *Burden*: Will it address the most pressing needs?
– *Appropriateness*: Will it be affordable, robust, and adjustable to settings in developing countries, and will it be socially, culturally, and politically acceptable?
– *Feasibility*: Can it realistically be developed and deployed in a time frame of ten years?
– *Knowledge gap*: Does the technology advance quality of life by creating new knowledge?
– *Indirect benefits*: Does it address issues such as capacity building and income generation that have indirect, positive effects on developing countries?

Not only did the study identify the top ten key applications, but it was also able to prioritize the importance of each application niche and define relevant examples (Table 1.7). Priorities were derived by having each panelist rank their top 10 choices from the 13 key applications selected on the previous round. With 63 panelists and 13 choices, the maximum score any given application could receive was 819.

The MDGS have been updated periodically over the intervening years and have driven considerable positive change, especially in developing countries. In 2017, they shifted direction, morphing from "Millennial" Development Goals" (MDGs) to "Sustainable" Development Goals" (SDGs) (Figure 1.10) [37].

As discussed in an ACS NanoFocus, "Nanotechnology for a Sustainable Future: Addressing Global Challenges with the International Network4Sustainable Nanotechnology" (Dec 2021), nanotechnology is seen a prime driver in realizing these UN goals [38].

They pose the question, "Why Nanotechnology?" First, as shown in Figure 1.11, Nanotechnology is central to a constellation of new technology development, such as AI, robotics, Internet of Things (IoT), and Smart Infrastructure.

Second is nanotech's connectivity to the key application challenges posed by the SDGs. Figure 1.12 ties a wide range of Nanotechnology application clusters to those goals, validating Nanotechnology's role in propelling human evolution to its next level, "Soci-

Table 1.7: Correlation between the top ten applications of Nanotechnology for developing countries and the UN millennium development goals (MDGs) (MDGs)'' of 2005. These goals illustrate the wide-ranging applications for Nanotechnology.

Ranking (Score)	Applications	Examples	Comparison with the MDGs
1 (766)*	Energy storage, production, and conversion	Novel hydrogen storage systems based on carbon nanotubes and other lightweight nanomaterials Photovoltaic cells and organic light-emitting devices based on quantum dots Carbon nanotubes in composite film coatings for solar cells Nanocatalysts for hydrogen generation Hybrid protein-polymer biomimetic membranes	VII
2 (706)	Agricultural productivity enhancement	Nanoporous zeolites for slow-release and efficient dosage of water and fertilizers for plants, and of nutrients and drugs for livestock Nanocapsules for herbicide delivery Nanosensors for soil quality and plant health monitoring Nanomagnets for removal of soil contaminants	I, IV, V, VII
3 (682)	Water treatment and remediation	Nanomembranes for water purification, desalination and detoxification Nanosensors for the detection of contaminants and pathogens Nanoporous zeolites, nanoporous polymers, and attapulgite clays for water purification Magnetic nanoparticles for water treatment and remediation TiO_2 nanoparticles for the catalytic degradation of water pollutants	I, IV, V, VII
4 (606)	Disease diagnosis and screening	Nanoliter systems (Lab-on-a-chip) Nanosensor arrays based on carbon nanotubes Quantum dots for disease diagnosis Antibody-dendrimer conjugates for diagnosis of HIV-1 and cancer Nanowire and nanobelt nanosensors for disease diagnosis	IV, V, VI
5 (558)	Drug delivery systems	Nanocapsules, liposomes, dendrimers, bucky balls, nanobiomagnets, and attapulgite clays for slow and sustained drug release systems	IV, V, VI
6 (472)	Food processing and storage	Nanocomposites for plastic film coatings used in food packaging Antimicrobial nanoemulsions for applications in decontamination of food equipment, packaging or food Nanotechnology-based antigen detecting biosensors for identification of pathogen contamination	I, IV, V

Table 1.7 (continued)

Ranking (Score)	Applications	Examples	Comparison with the MDGs
7 (410)	Air pollution and remediation	TiO_2 nanoparticle-based photocatalytic degradation of air pollutants in self-cleaning systems Nanocataysts for more efficient, cheaper, and better-controlled catalytic converters Nanosensors for detection of toxic materials and leaks Gas separation nanodevices	IV, V, VII
8 (366)	Construction	Nanomolecular structures to make asphalt and concrete more robust to water seepage Heat-resistant nanomaterials to block ultraviolet and infrared radiation Nanomaterials for cheaper and durable housing, surfaces, coatings, glues, concrete, and heat and light exclusion Self-cleaning surfaces (e. g., windows, mirrors, toilets) with bioactive coatings	VII
9 (321)	Health monitoring	Nanotubes and nanoparticles for glucose, CO_2, and cholesterol sensors and for in-situ monitoring of homeostasis	IV, V, VI
10 (258)	Vector and pest detection and control	Nanosensors for pest detection Nanoparticles for new pesticides, insecticides, and insect repellents	IV, V, VI

Figure 1.10: In 2017, the UN's "Millennial Development Goals" (MDGs) have been expanded into "Sustainable Development Goals" (SDGs) (SDGs)'' with an achievement target of 2030.

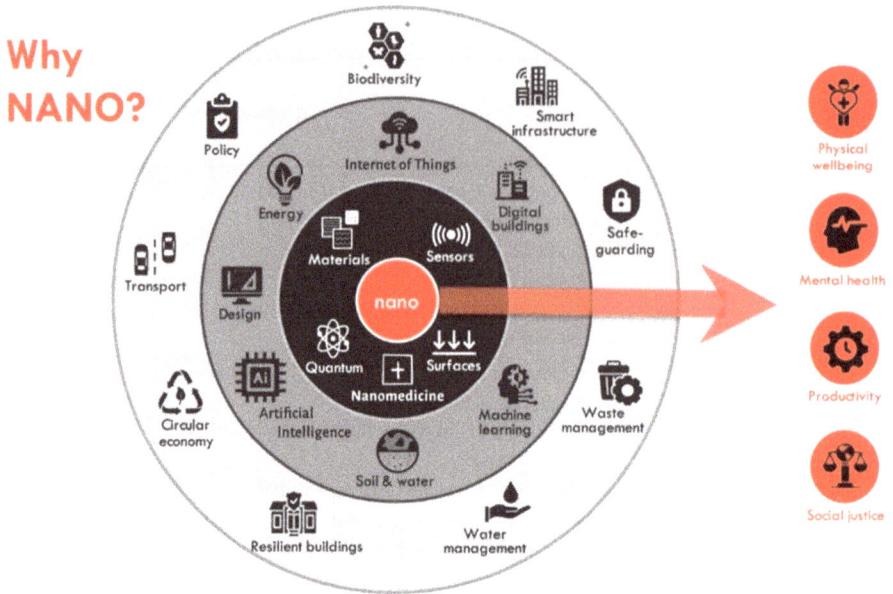

Figure 1.11: Why Nanotechnology is central to developing all the applications needed to reach the UN's Sustainable Development Goals. ACS Nano 2021, 15, 12, 18608–18623.

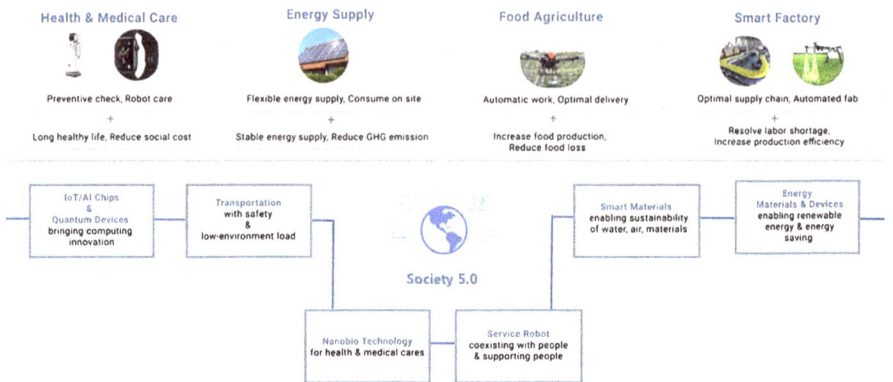

Figure 1.12: How specific nanotechnology application clusters answer the needs of "Society 5.0". (Society 1.0 = hunter-gather; 2.0 = agriculture; 3.0 = industry; 4.0 = information; 5.0 = knowledge/human-centric/sustainable). ACS Nano 2021, 15, 12, 18608–18623.

ety 5.0". This level is knowledge rather than just information-based, is human centric (leaving no one behind), and is sustainability focused.

Because of these two factors, Nanotech enjoys extensive and far-reaching participation from government, industry, and academe and broad-based funding from diverse sources, allowing it to flourish and progress.

This article was also helpful, because it went beyond a discussion of applications into the process of commercialization, bringing into focus the importance of life-cycle planning and nanosafety and the interplay of government, academe, and the private sector. At this point, however, this author encourages caution. Whereas the ACS authors bring up very valuable points, their discussion tends to be more academic, suggesting limited experience with the practicalities of real-life commercialization. For example, they cite the importance of life cycle planning and safety. Having been in technical marketing and strategic planning roles, working in and consulting to dozens of companies for over 40 years, this author has never found an instance in the real-world in which a company did not take these two issues into consideration when moving from RD&E to commercialization. Secondly, the ACS authors stressed the importance of "circularity": of taking ideas from the lab to real-world then getting feed-back from customers to drive the next iteration of technology. Again, with over four decades experience, this author found that most companies valued input from customers and implemented formalized processes, such as "users groups" to collect and organize information to be fed back into RD&E. One point that the ACS group mentioned that this author would like to underscore is the importance of getting to know your audience BEFORE you build technology. This is a point on which I actually built my career. Early-stage user-based research is, indeed, critical. Not only can it save a company from going down a very expensive wrong way street, it can shorten the time to commercialization and guarantee salability, as explained in this intriguing article, "Who is going to buy the darn thing?" [39].

1.8 The Nanotechnology Market

All of the previous discussions illustrate Nanotechnology's heterogeneity. Therefore, both defining the market and sizing it present challenges. To begin with, there is some debate as to whether Nanotechnology is really a product-based industry or whether it is "simply, a set of enabling technologies that supports many existing industries (basically applying the 'nano' label to existing technologies: electronics, optics, composite materials, pharmaceuticals, etc.)." [40]

1.8.1 Is There Really a "Nanotechnology Market"?

The debate began with Cientifica's "Nanotechnology opportunity report" in 2003, which stated that "what exists. . . is a teeming collection of technologies and applications seeking each other." [41] About the same time, Tim Harper, one the report's authors, reiterated his position that Nanotechnology is still not really an industry [42]. However, in December 2020, Michail Rocoa of the National Science Foundation (NSF) and the National Nanotechnology Initiative (NNI) definitively stated that "Nanotechnology is a foundational, general-purpose [Science and Engineering] field… It is a global science initiative." [43]

When material was collected for this chapter, considerable evidence supported the position that Nanotechnology was more than just an enabling technology or a "teeming collection of technologies and applications." First, consider the rapid proliferation of technical journals and conferences dedicated specifically to Nanotechnology; its subsets also lends credence to its existence as a separate industrial sector. StatNano reports that there are now over 2,000,000 articles on nanotechnology in the Web of Science (WoS). Table 1.8 shows the annual world-wide growth in published articles from 2018 to the present.

Table 1.8: StatNano.com reports over 2,000,000 nanotechnology articles are posted on the Web of Science (WoS). This table shows the world-wide output, by year, from 2019 through *June 2023 [44]. (Note: StatNano.com offers detail from 68 countries in a searchable spreadsheet).

World	2019	2020	2021	2022	2023*
# articles/yr	179,664	197,582	212,478	234,169	112,072

Geographically, there is wide participation in the nanolandscape. As mentioned in the caption for Table 1.8, StatNano.com provides searchable database of 68 countries for the articles listed. Not only are companies and publications proliferating around the globe, universities with Nanotechnology programs are sprouting up everywhere, often appearing in unexpected locations, including Africa (Figure 1.13).

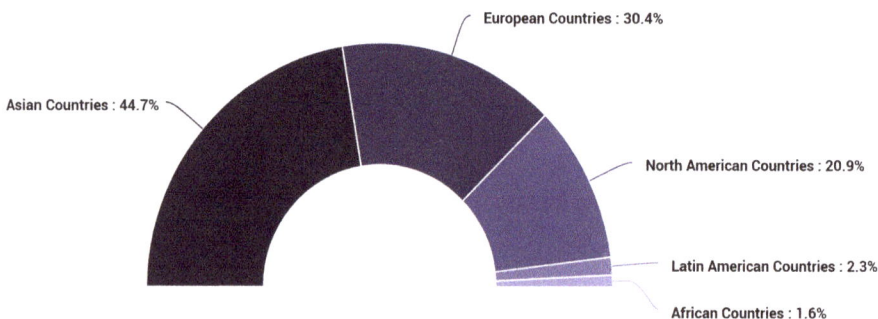

Figure 1.13: Global distribution of Universities with Nanotechnology programs. Asia (363), Europe (247), North America (170), Latin America (19), Africa (13) [45].

Next, consider the number of Nanotechnology patents (Figure 1.14). StatNano reports that there are now over 175,000 Nanotechnology patents registered with the USPTO [46].

Nanotechnology published patent applications in USPTO:
Number and Annual Growth Rate during the Past 20 Years

StatNano

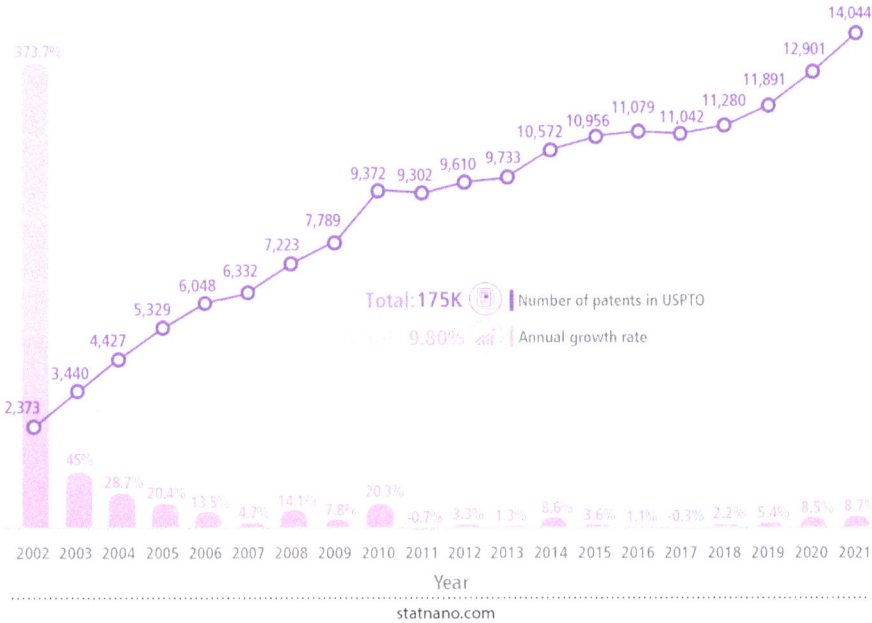

Figure 1.14: Growth in number of patents, 2002 through 2021 shows a total of 175,000 nano patents filed in the USPTO, with an annual growth rate of nearly 10 % [47].

It is also clear that professionals in both research and development place significant value on Nanotechnology. The 2022 R&D Global Funding Forecast [48] surveyed the "Most important technologies by 2025." As shown in Figure 1.15, though artificial intelligence dominates, Nanotechnology placed strongly in second tier of technologies.

Also, consider the proliferation of products. In 2005, PEN established the Consumer Product Inventory [49]. It featured only 1800 products. A more current list, available at https://product.statnano.com/ now boasts over 10,821 products from over 3643 companies located in 68 countries. Both StatNano.com and Nanowerk.com provide global roll country-by-country (state-by-state for the USA) that includes both commercial and research nanoactivities.

Table 1.9 summarizes the nanoactivities from NanoWerk. It compares data originally reported in this chapter in 2016 to the current listing. A third column was added to augment with data from the larger StatNano database, where available. Since the number of universities and products have skyrocketed since 2016, it is unclear whether any consolidations seen since 2016 suggest a maturing of this market or underreporting.

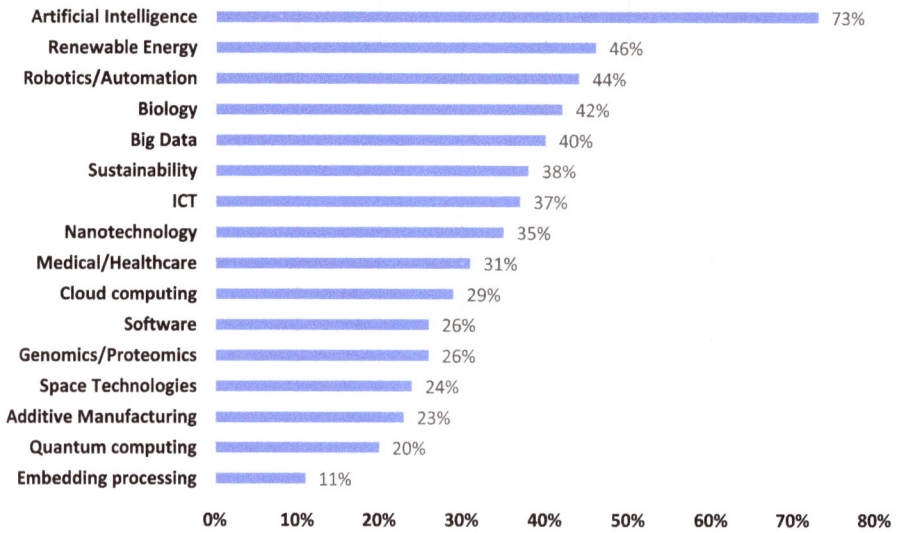

Figure 1.15: "Most important technologies by 2025", 2022 R&D Global Funding Forecast. www.RDWorldonline.com (Data has been reformatted from the original alphabetical listing to a relative ranking).

Table 1.9: Comparison of data from NanoWerk.com, 2016 to 2023. Augmented with findings from Stat-Nano.

	2016 data	2023 data	StatNano
Nanomaterials	>2500		
Nanotechnology companies and laboratories	2057	1334	5410
Universities departments, labs, and research groups	131	1176	847
Government, industry, and private labs	291	235	
Initiatives and networks (national & international)	430	303	
Associations and societies	25	18	
Services, intermediaries, and "other"	197	98	
Raw materials	292	284	
Biomed and life sciences	275	176	
Products, apps, instruments, and technologies	1208	6380[*]	10,860[*]
Science and degree programs (total)[*]	301	273	
BSc	67	66	
MSc	155	145	
PhD	48	38	
Other – certifications, etc.	31	24	
Conferences & events	>100	20	

[*] Number of products shown in their "Products" listing

1.8.2 What Is the Size of the Nanotechnology Market?

The Nanotechnology market is highly heterogeneous (Figure 1.16). As a result, it is difficult to size accurately. The discussion below highlights both data and some of the pitfalls encountered when reading off the shelf market research in this arena.

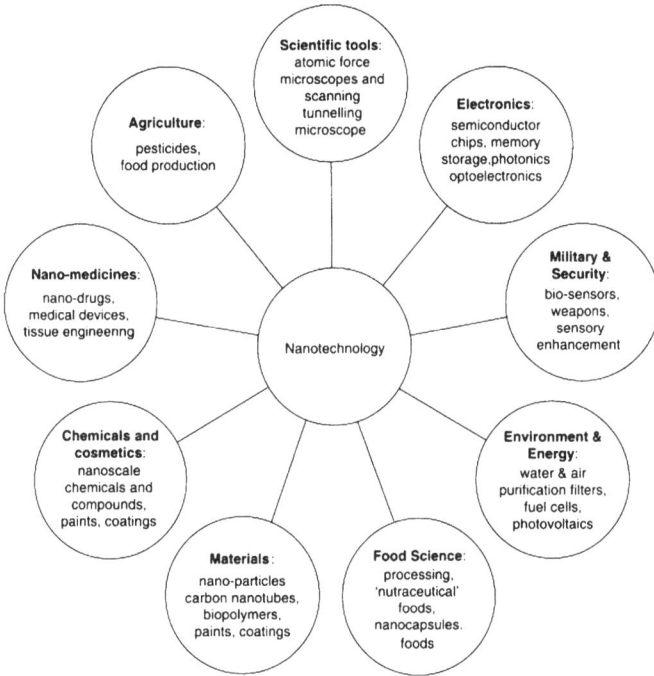

Figure 1.16: Nanotechnology is a highly heterogeneous discipline [50].

In 2013, NSF and the *National Nanotechnology Coordination Office* (*NNCO*) funded an independent study to determine the global revenue from nano-enabled products. The results were published in the LUX Research Report, an extract of which follows:

Governments, corporations, and private investors (venture capitalists) invested 18.5 billion in Nanotechnology in 2012, increasing their spending 8 % relative to 2010. The US contributed 36 % of this amount. Corporations expanded spending by 21 % over 2010, whereas governments and private investors reduced their investments by 5 % and 10 %, respectively. The United States maintained its lead over all other governments, with $2.1 billion of federal and state funding in 2012. US corporations also led global spending on Nanotechnology research and development, investing $4 billion in 2012, which was approximately $1 billion more than the next country, Japan. The revenue from nano-enabled products has continued to grow, from $339 billion in 2010 to $731 billion in 2012. This total is a slight decrease in our estimate relative to our last update on nano-enabled product revenues released in 2009. Our expanded forecast for nano-enabled products reveals the global value of nano-enabled products, nano-intermediates, and nanomaterials reaching $4.4 trillion by 2018 [51].

The accompanying press release from NSF reported that the survey "shows global funding for emerging Nanotechnology has increased by 40-to-45 percent per year for the last three years [2010–2013]." [52]

Data from this report has been quoted extensively, but there is a major flaw in the presentation of these findings, which has triggered extensive controversy. The report cites "nano-enabled" technology and "nano-intermediates." Why are these terms misleading? Consider the case of a car that has a nanocoating. The nanomaterial forms the coating on the car and should be counted in the category "nanomaterials." It would have a value of a few dollars. However, this report tallies the *nano-enabled technology* as the entire car, valued at thousands of dollars. A more reasonable approach is to size the market on the actual Nanotechnology itself.

1.8.2.1 BCC Research, "Nanotechnology: A Realistic Market Assessment"

In 2013, BCC released "Nanotechnology: A Realistic Market Assessment." For the purposes of this report, BCC defined Nanotechnology applications as "the creation and use of materials, devices, and systems through the manipulation of matter at scales of less than 100 nm."

[The report included] nanomaterials (nanoparticles, nanotubes, nanostructured materials, and nanocomposites), nanotools (nanolithography tools and scanning probe microscopes), and nanodevices (nanosensors and nanoelectronics).

[The report did not include nanoscale semiconductors, but did include] the tools used to create them. [It also excluded very high volume, materials that have been used long before nanotechnology came into existence such as] carbon black materials carbon black nanoparticles used to reinforce tires and other rubber products; photographic silver and dye nanoparticles; and activated carbon used for water filtration [because their data would] tend to swamp the newer nanomaterials in the analysis."

As seen in Table 1.10, the BCC market estimates have cooled considerably. One reason may be that data is now being split out into different market segments and reports.

Table 1.10: Comparison of data from several recent market studies.

	Begin	$	Projected	$	CAGR Per.	%
BCC 2014 Report [53]	2014	$22.9B	2019	$64.2B	2014–2016	19.8 %
BCC 2020 Report [54]	2021	$5.2B	2026	$23.6B	2021–2026	35.5 %
North America	2021	$1.6B	2026	$7.2B	2021–2026	34.5 %
Asia-Pacific	2021	$1.2B	2026	$6.0	2021–2026	37.6 %

For instance, when Google was searched for "BCC Nanotechnology Market," it pointed to a BCC page listing 61 Nanotechnology reports!

Clearly, the results from one report to another vary dramatically. What guidelines can be used to best interpret these results? The critical caveat is "read carefully." The following questions should be asked:

- Is the report citing funding or revenue? The Lux report, for example, reports both.
- Is it discussing the actual nanocomponent/technology or the sales of products containing or *enabled* by Nanotechnology? Calculations for the LUX report are based on nano-enabled products; these report on the Nanotechnology itself.
- Which market sectors are included and which are not? For example, in discussing the global Nanotechnology market, the BCC report does not include semiconductor manufacture, but does include the instrumentation used for semiconductor testing and measurement.
- Finally, what is the margin of error? The author has encountered reports that cited imprecisely presented assessments, such as "this trend represents 4 % of the market ±10 %". In other words, that particular trend could be anything from 0 % to 14 %. Be cautious when assessing actual numbers.

An interesting case in point were data reported by NNI's Mikhail Roco. In his December 2020 presentation (cited above, Slide #6), he suggests that the market will rise to $30Trillion in new products by 2030, a thousand-fold difference! Since he is an "insider," working within the biggest Nanotechnology initiative in the world, one would think that there is some validity to his estimate, but why such a huge difference? One answer might be "NBICA," a powerful new driver in the Nanotechnology arena. NBICA is the acronym for the convergence of Nanotechnology, Biology, IT, Computation, and AI. Roco points out that this convergence of technologies is prompting an "innovation spiral [leading to] new Nanotechnology architectures; spin-off disciplines and technology platforms; new applications, sectors & businesses; and new expertise." StatNano echoes his findings and have opened a special news section called NBIC+, which reports news from multiple perspectives, including Industry & Market, Research & Science, Education, Policy, Safety, and Society [55]. They point out "hence, technology is on the verge of a significant change in its path: a strategic shift from divergence to convergence, from specialization to synergy, which undoubtedly will go down in the history of science and technology."

1.8.3 Nanotechnology Market Sectors

As with applications, the reader is encouraged to research market sectors of interest. Select data are included below for several of the more relevant sectors.

1.8.3.1 "Graphene, 2-D materials, and Carbon Nanotubes

This paragraph from Universal Matter (https://www.universalmatter.com/) explains why graphene is so important: "Graphene has the potential to provide a significant environmental advantage to the human race by strengthening, and therefore reducing the volume used, of many of the raw materials we use every day: concrete, asphalt, plastics, rubber and paint to name just a few. For applications where the volume cannot be reduced, graphene promises to significantly improve the life cycle of products. For example graphene in asphalt has been shown to increase the life of a road by 300 %."

According to IDTechEx, the market will be segmented across many applications, reflecting the diverse proper ties of graphene, and project that the market for functional inks and coatings will make up 21 % of the market by 2018. Ultimately however, energy storage and composites will grow to be the largest sectors, controlling 25 % and 40 % of the market in 2026, respectively." [56]. Figure 1.17, taken from Grandview research, depicts their projections for the graphene platelets, graphene oxide, and reduced oxide from 2018 through 2028.

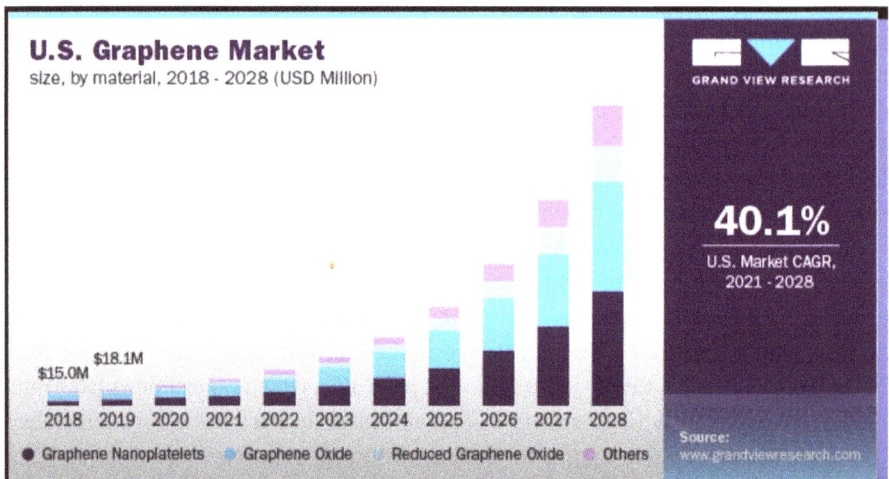

Figure 1.17: Ten-year market projections split by application. Inset: market share of graphene platelets, oxide, reduced oxide and "others from 2018–2028, USD Million. Source: Grandviewresearch.com.

When this book was originally written, graphene promised to be Nanotechnology's wonderchild. Yet, nearly 20 years after its discovery, it has produced only moderate results. A YouTube video, "Why hasn't graphene taken over the world... yet?" [57] suggests three key the reasons: cost and inability to produce commercial quantities at sufficient purity.

Both new methods for producing graphene and new applications are coming on line with great regularity. While still limited by cost and consistent purity, new production methods, including the special graphene used by GAC in their ultra-fast charging graphene battery (GAC) [58] and Flash Graphene (Universal Matter) [59], are making large batch graphene a reality.

It will be interesting over the next seven years to see how these new approaches change the reality of the graphene market.

1.8.3.2 Nanotechnology Drug Delivery Market

In 2015, Transparency Market Research issued a report specifically targeted to drug delivery. "Advancement in nanotechnology has revolutionized the delivery of nanometer-range drug molecules. Rising prevalence of infectious diseases and cancer, significant nanotechnology research, and increasing demand for novel drug delivery systems are driving the nanotechnology market. However, unspecific regulatory guide-lines for novel nanotechnology-based drugs are expected to hamper market growth during the forecast period."

That earlier report suggested that the global Nanotechnology drug delivery market was valued at $4.1 billion in 2014 and was anticipated to reach $11.9 billion by 2023, expanding at a CAGR of 12.5 % from 2015 to 2023." [60] More current findings from Precedence Research (Figure 1.18) states "The global nanotechnology drug delivery market size was reached at USD 83.8 billion in 2022 and is expected to hit around USD 183.11 billion by 2032, poised to grow at a CAGR of 8.13 % between 2023 and 2032." [61]

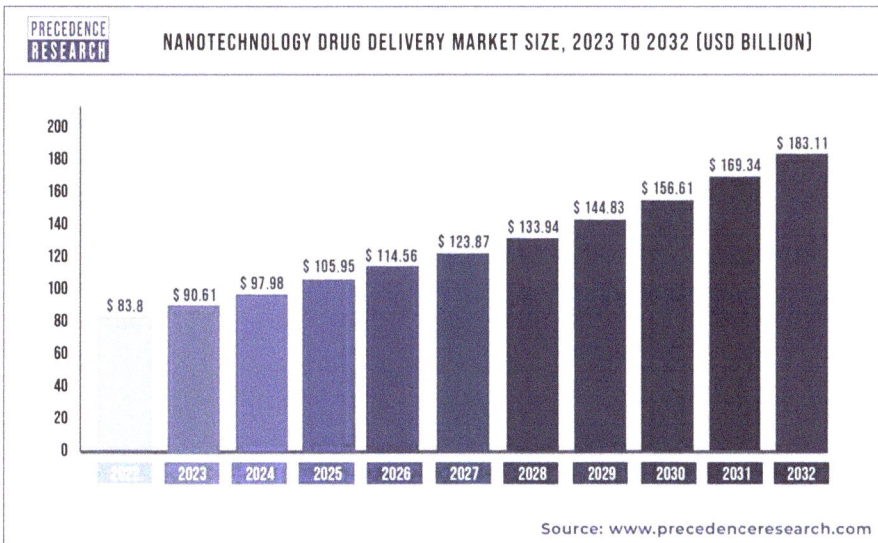

Figure 1.18: Nanodelivery of drugs gained prominence during the COVID crises, driving demand at a projected CAGR of 8.13 % through 2032. (Precedence Research).

1.8.3.3 2021–2025 Food

Food security is a high priority world-wide is one of the UN's Sustainable Development Goals. New applications of Nanotechnology in water management, farming, and fertilization use promise increased yields for a growing global population. This report from TechnavioPlus [62] suggests that the food Nanotechnology market share is expected to increase by USD 187.84 billion from 2020 to 2025, and the market's growth momentum will accelerate at a CAGR of 25.32 %.

1.8.3.4 Wearables and Smart Textiles

Cientifica's report on Wearables [63] gives an intriguing glimpse of the future. "Unlike today's 'wearables,' tomorrow's devices will be fully integrated into the garment through the use of conductive fibers, multilayer 3-D printed structures and two-dimensional materials such as graphene. [Nanotechnology will be involved] from antibacterial silver nanoparticles to electrospun graphene fibers [… and will] impact on sectors including wearables, apparel, home, military, technical, and medical textiles.

"As shown in Figure 1.19, Global Industry Analysts, Inc." (www.StrategyR.com) projects a major jump in nanotextiles. "The global market for Smart Textiles estimated at US$2.3 Billion in the year 2022, is projected to reach a revised size of US$13.8 Billion by 2030, growing at a CAGR of 24.9 % over the analysis period 2022–2030." [64]

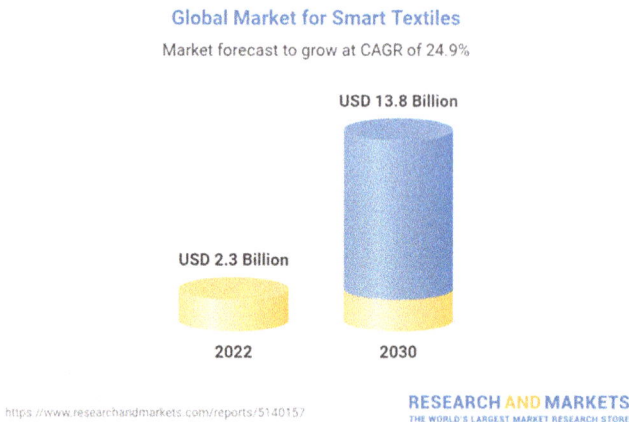

Figure 1.19: The global market for nano-enabled smart textiles is expected to jump an energetic 24.9 CAGR. (Global Industry Analysts, Inc." www.StrategyR.com).

1.8.3.5 Internet of Nanothings (IoNT)

"The Internet of Nano Things Market consists of nano things that are connected to the Internet, that is, anything, anytime, and anywhere. It consists of integrating sensors and devices into everyday objects that are connected to the Internet over fixed and wireless networks. In the first printing of this book, the Internet of Nano Things (IoNT) Market was expected to grow from \$4.26 billion in 2016 to \$9.69 billion by 2020, at an estimated compound annual growth rate (CAGR) of 22.81 % from 2016 to 2020." [65]

As shown in Figure 1.20, "Prudour anticipates this market segment to reach USD \$123.5 Bn by 2032 from USD \$16.4 Bn in 2022, rising at a compound annual growth rate (CAGR) of 22.38 % between 2023 and 2032." [66]

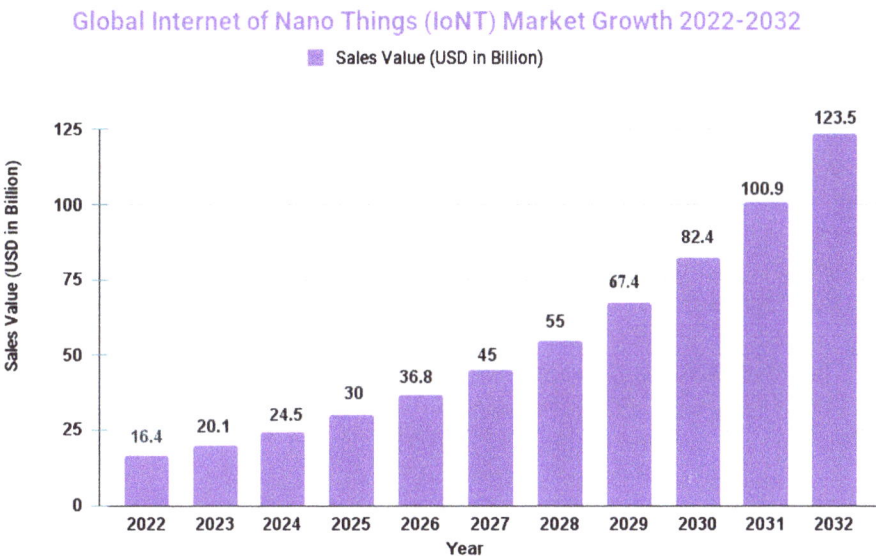

Figure 1.20: With an estimated CAGR of 22.38 % over the next decade, the Internet of NanoThings is one of the fastest growing segments of the market. (Prudour).

1.9 The Challenge of Nanotechnology Safety

As discussed at the beginning of this chapter, Nanotechnology, like radiation, presents unique safety challenges, because you cannot see it, cannot smell it, cannot taste it, cannot feel it, and cannot hear it. But, unlike radiation, which comes in relatively few forms, Nanotechnology presents dangers from an ever-growing panoply of different structures, forms, and chemical constitutions. The issue of open-ended versus closed CNTs was discussed in Section 1.3.1. In QDs, other hazards come from the materials in the semiconductor. Cadmium is a prime example. In the biological realm, the threat could come from the mutation of a virus used as a carrier. Each sector and material has its own risks and hazards, from air-borne particles to materials that are toxic on contact or through ingestion.

Even more than with radiation, there are many unknowns with far-reaching implications. That topic is addressed in greater detail by Walt Trybula and Deb Newberry in Chapter 6 of this book. Another challenge involves assessing which information is truly reliable, as discussed by Evelyn Hirt and Walt Trybula in Chapter 7. For example, in writing this introductory chapter, a great deal of material surfaced from the 2005–2008 time period. The author had to weigh carefully whether concepts from that era still applied: Technology and tools have evolved considerably over the intervening decade. Similarly, the market analyses included in this chapter required a keen eye to assess which information was valid, which presented only part of a story, and which was overly enthusiastic. Those of you interested in pursuing nano-safety as a career will need to be well-grounded in science, technology, engineering, and math (STEM), but also exercise a healthy skepticism and organization of thought.

Clearly, nano-safety is very much on the mind of the US government. At a first level, multiple organizations, including OSHA, NIOSH, EPA, FDA, and CDC, are beginning to compile standards and methods that will be helpful in meeting these challenges. Second, through NNI, the government is establishing foundational research to focus specifically on Nanotechnology-related environmental health and safety (EHS). In 2011, NNI developed the EHS Research Strategy, with a charter to coordinate, map, and implement all governmental EHS efforts in Nanotechnology. The latest progress review of that group (2017) [67] outlines six core research areas, which provide the following framework:
- Nanomaterial measurement infrastructure
- Human exposure assessment
- Human health; Environment
- Risk assessment and risk management methods
- Informatics and modeling

Industry has long been on-board, with companies adding EHS officers early in the commercialization process [68].

1.10 The Crucial Need for Education and Certification

In April 2016, NNI's working group on Nanotechnology Environmental and Health Implications (NEHI) launched the webinar, "Applying a lab safety culture to nanotechnology: Educating the next generation." [69] The presentation summarizes both the need for a culture of safety in all areas of nanotechnology and for proper education and certification in this field. NEHI [70] encompasses 19 government agencies and collaborates extensively with industry and academia, both within the USA and abroad.

Kicking off the webinar, panelist Larry Gibbs, Associate Vice Provost for EHS at Stanford University, discusses the "laboratory safety culture spectrum" (Figure 1.21), outlining the hierarchy of attitudes about safety, from the lackadaisical "pathological" to the attentive, proactive, culture-based "generative." An important point to take away: if safety consciousness and education are not managed properly and consistently, it is very easy for a facility to slide down this scale.

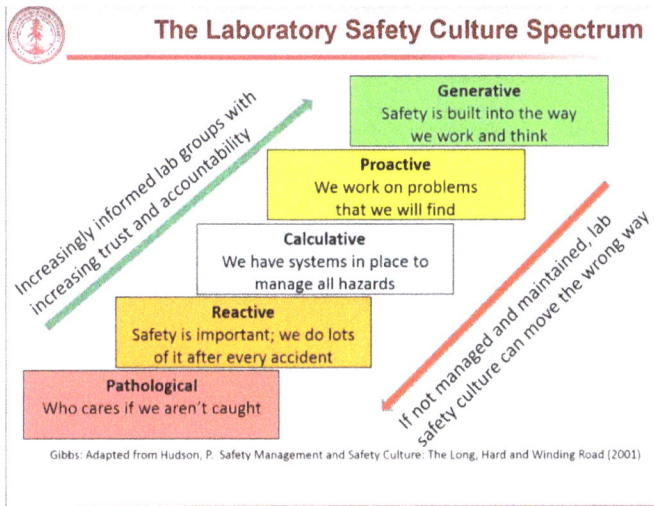

The Laboratory Safety Culture Spectrum

Generative
Safety is built into the way
we work and think

Proactive
We work on problems
that we will find

Calculative
We have systems in place to
manage all hazards

Reactive
Safety is important; we do lots
of it after every accident

Pathological
Who cares if we aren't caught

Increasingly informed lab groups with
increasing trust and accountability

If not managed and maintained, lab
safety culture can move the wrong way

Gibbs: Adapted from Hudson, P. Safety Management and Safety Culture: The Long, Hard and Winding Road (2001)

Figure 1.21: Laboratory safety culture spectrum. (From the webinar "Applying a lab safety culture to nanotechnology: Educating the next generation").

The second speaker, Dr Craig Merlic, Executive Director of the UC Center for Laboratory Safety and Associate Professor in UCLA's Department of Chemistry and Biochemistry, defines the "safety triad" (Figure 1.22), based on an institution's attitudes and commitment to safety, their formal protocols and procedures, and the staff's willingness to report honestly so that facility can learn from accidents and errors.

The last presenter is Lori Seiler, Global R&D EHS at Dow Chemical, who describes The Dow Lab Safety Academy (http://safety.dow.com), an online program developed in 2013. Since its inception, Academy lessons have been seen more than 250,000 times by more than 25,000 enrolled viewers. They have also provided services to >60 universities, >40 government agencies and national laboratories, and >100 companies in 10 countries.

At the Academy's core are four safety modules: (1) safety orientation and training; (2) specialized training; (3) how to plan, evaluate, and execute a safety culture; and (4) how to build a sustainable safety culture. Through these modules, the program trains the visitor in pretask planning, hazard awareness, how to build procedures and protocols,

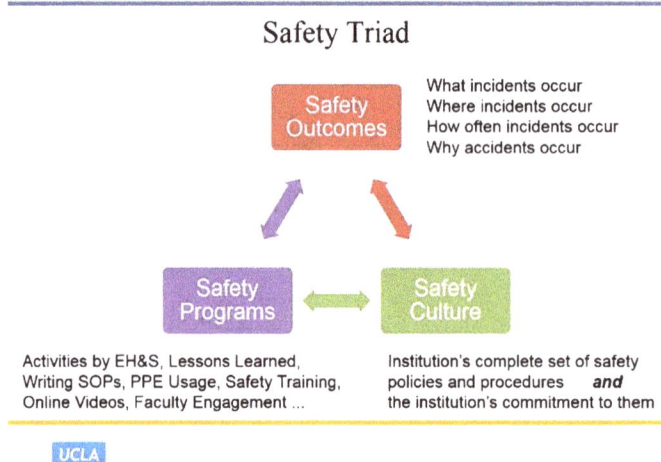

Figure 1.22: The safety triad, demonstrating the impact of safety programs and cultures on safety outcomes. (From the webinar "Applying a lab safety culture to nanotechnology: Educating the next generation").

emergency planning, and the importance of leadership engagement, accountability, and ownership. All modules reinforce the message, "You are your own best safety advocate."

A number of programs have answered the call for education, training, and certification. An important resource is NanoHub (www.nano.gov/education-training), a searchable database of Nanotechnology education resources.

One of the most extensive programs in the USA is Nano-Link (www.atecentral.net/r8287/nano-link_center_for_nanotechnology_education). This NSF-funded project brings together 11 institutions throughout the Mid-West that provide workforce development, teacher training, and classroom resources. They also partner with local industries "to promote nanotechnology education at multiple grade levels by providing comprehensive resources for students and educators ... supported by hands-on educator workshops and online content and activity kits." Nano-Link organizer, Deb Newbury, has also posted the following two key presentations on the website: [71]

- NCPN (National Career Pathways Network) 2012: A correlation study between emerging technology concepts and job requirements
- NSTA 2015: What is Nano-Link?

METPHAST [72] (Midwest Emerging Technologies Public Health and Safety Training) is also part of Nano-Link. Funded by the National Institute of Environmental Health Sciences Superfund Research Program, METPHAST is a multi-institutional collaboration between the University of Minnesota/School of Public Health, the University of Iowa College of Public Health, and Dakota County Technical College. The group has constructed 20 web-based modules on nano-safety, which are available for public use. The International Association of Nanotechnology Training Programs (IANT) [73] also provides professional courses and certification in both nano-safety and clean technology.

Overseas, Oxford University also has a series of both degree programs and short courses in Nanotechnology and Nanomedicine [74].

Both NanoWerk.com and StatNano.com offer listings of universities world-wide. NanoWerk cites 273 degree programs in Nanotechnology fields but goes on to say, "Chemists, physicists, biologists, materials scientists – they all view Nanotechnology as a branch of their own subject. And this view is reflected in the smörgåsbord of offerings: Bachelor and Masters programs in Micro- and Nanotechnology; Nanomedicine; Nanotechnology and Microfabrication; Nanoscience; Micro- and Nanosystems; Nanobiosciences and Nanomedicine; Nanobiology; Nanoengineering; Photonics Engineering, Nanophotonics and Biophotonics; etc.

Less formal courses are also flourishing, especially in Europe. One is United Nations Institute for Training and Research (Unitar) Nanomaterials Safety Course [75]. This 5 day, web-based course is offered at no cost.

Here is a partial listing of other opportunities:

– SUSNANOFAB – Nanosafety Training International Iberian National Laboratory (INL) 03/27/2023 (https://youtu.be/v1Sll3J1ous)
– Nanosafety Training School 2023, May 015-19, Venice Italy: SSbD Approaches for Chemicals, Advanced Materials and Plastics (#venicenano23)
– NanoSafety -Lund University (https://luvit.education.lu.se/luce/activities/activitydetails_ext.aspx?id=1639)

For more formalized degree programs, there are a number of resources, including NNI (www.nano.gov), which provides the US Technology Resource, an interactive map that shows the locations of higher education degree programs, and a link to Penn State's Nanotechnology Applications and Career Knowledge Center (NACK) for the development of community college programs. Also, visit https://www.nanowerk.com/ and Statnano.com.

Although all of these programs feed into professional education and certification in Nanotechnology, the need for education begins earlier and more broadly, with STEM (educational programs in science, technology, engineering, and math) and in community outreach. For STEM, Nano-Infusion (another component of Nano-Link) offers a collection of modules, including 10 min demos and 20 or 50 minute experiments. They provide the supplies (no charge), training, and support, as well as a "nano-geek" T-shirt. Important components of the package are excellent online videos that demonstrate each activity.

In support of career education, Arizona University's Nanotechnology Cluster K-20 provides Nanotechnology scientists, who educate students at all levels on "intricacies and careers in nanotechnology." Everyone can learn more about Nanotechnology during NanoDays [76], a nationwide week-long festival held each year during the last week in March/first week in April at more than 250 science museums, children's museums, research centers, and universities.

1.11 The Future

The future of Nanotechnology is limited only by our imaginations. The NNI has a special, evolving list of Nanotechnology signature initiatives (NSIs) that "spotlight topical areas that exhibit particular promise, existing effort, and significant opportunity, and that bridge across multiple Federal agencies." [77]

1.11.1 NNI's Signature Initiatives (as of 2021)

Here is the current list and the charter for each initiative.

Goal 1. Ensure that the United States remains a world leader in nanotechnology research and development

At the heart of the NNI is support for Nanotechnology R&D across the entire continuum, from basic research that fuels new discoveries through application-driven advanced research and development that leads to new products in the market. Proposed solutions leveraging the unique properties at the nanoscale successfully compete for funding in general research solicitations, and agencies now rarely specify Nanotechnology in requests for proposals, in contrast to the early days of the NNI. While this reflects the successful embedding of Nanotechnology throughout the R&D enterprise, it makes coordination more difficult, and perhaps more important than ever to ensure that resources, knowledge, and synergies are fully leveraged. The NNI agencies will continue to use their full suite of authorities and mechanisms to fund Nanotechnology R&D, and more deliberate mechanisms will be used to connect and build communities, both within the NNI and with other initiatives and priorities. National Nanotechnology Challenges are being introduced in this plan to mobilize the Nanotechnology community to help address global issues.

Goal 2. Promote commercialization of nanotechnology R&D

The NNI will enhance efforts to accelerate the scale-up, translation, and commercial application of Nanotechnology R&D into the marketplace to ensure that economic, environmental, and societal benefits are realized and to help the Nation build back better with high-paying jobs. Additional efforts supporting this goal are perhaps the most significant updates to the strategic plan and address the critical needs identified by the NASEM committee. More explicit connections will be made to broad agency efforts that support transition of nanotechnologies to the regional ecosystems that exist across all of America. Efforts will include expanding the Nanotechnology Entrepreneurship Network6 as a forum to connect innovators and share best practices.

Goal 3. Provide the infrastructure to sustainably support nanotechnology research, development, and deployment

The need for expensive, specialized tools remains a key requirement for much of Nanotechnology R&D. A hallmark of the NNI has been the support of physical and cyber user facilities, which not only democratize nanoscience across the ecosystem but also serve as a platform to educate and train students who will become the next generation of instrument users, designers, and builders. NNI agencies recognize the importance of providing access to cutting-edge tools by upgrading and refreshing toolsets, as well as supporting the acquisition and maintenance of "workhorse" instruments. The NNI will support the increasing role of the cyber infrastructure (e. g., models, simulations, and data) that is critical for Nanotechnology innovation enhanced by artificial intelligence, machine learning, and advanced design tools. Facilities that support prototyping and early stages of the manufacturing process are also important for the development community and will be explored in collaboration with the private sector.

Goal 4. Engage the public and expand the nanotechnology workforce

The future of the NNI depends on a highly skilled workforce across the entire technology development pathway. In addition to targeted Nanotechnology education, the novel properties at the nanoscale can provide a spark to excite students to pursue science, technology, engineering, and mathematics (STEM) careers and help build a robust domestic workforce. NNCO and the NNI agencies use a variety of mechanisms to support public outreach and education from "K to grey" and will emphasize opportunities and access to resources, especially for people in traditionally underserved communities. In recognition of the importance of education, workforce development, and public engagement to the entire Nanotechnology enterprise, these areas are now a stand-alone goal of the NNI.

Goal 5. Ensure the responsible development of nanotechnology

The responsible development of Nanotechnology has been an integral part of the NNI since its inception, and the initiative has proactively considered potential implications and technology applications at the same time. Just as scientific understanding of nanomaterials has deepened and matured, the understanding of responsible development also has evolved. The responsible development framework articulated in this plan embraces new ideas that have emerged and builds upon concepts originally included in the NNI's responsible development efforts. A key tenet of responsible development remains the protection of human health and the environment through an understanding of not only the applications of nanomaterials, but also the potential implications. Responsible development further includes consideration of ethical, legal, and societal implications (ELSI) as well as a new emphasis on inclusion, diversity, equity, and access (IDEA) and the responsible conduct of research.

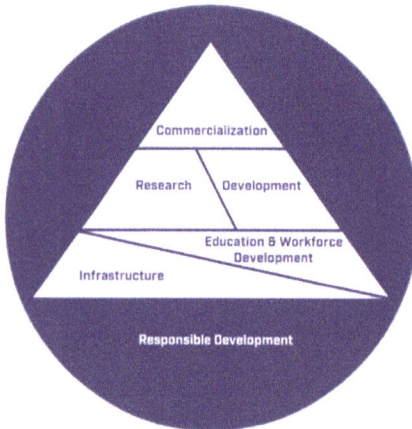

Figure 1.23: Diagram depicting the NNI goals. Credit: NNCO.

These five goals are highly interrelated and interdependent (Figure 1.23). World-class research requires top STEM talent, but also serves as a training ground for future researchers. Cutting edge equipment is required to conduct research and to attract top talent, but also supports the transformation of ideas into products, and enables education and workforce training efforts. The overarching principles of responsible development apply to all aspects of the NNI.

1.11.2 Other Intriguing Future Initiatives

On the commercial front, here are some thought-provoking excerpts from future research projects actually in progress.

1.11.2.1 Origami Robots

"Ido Bachelet of Bar-Ilan University in Israel and colleagues describe a set of 'DNA origami robots' that, when mixed in various combinations, function as biological logic circuits, capable of responding to the presence or absence of various molecular triggers.

"These 'robots,' aren't robots in the sense of Wall-E. They're DNA molecules that fold into complex and intricate shapes. Upon receiving the proper signals, they change shape, producing a measurable output.

"The robots were made in several forms. Some are 'effectors' capable of releasing a 'payload,' while others act as positive or negative regulators that modify the effectors' behavior. When combined in specific stoichiometric ratios and combinations, these robots can mimic the behavior of any of seven discrete 'logic gates': AND, OR, XOR, NAND, NOT, CNOT, and a 'half-adder.'" [78]

1.11.2.2 Programmable Food

"Researchers at US food giant Kraft have developed a colorless, tasteless liquid in the lab that consumers will design after you've bought it. You'll decide what color and flavor you'd like the drink to be, and what nutrients it will have in it, once you get home. You'll zap the product with a correctly tuned microwave transmitter. This will activate nano-capsules – each one about 2000 times smaller than the width of a hair – containing the necessary chemicals for your choice of drink: green-hued, blackcurrant-flavoured with a touch of caffeine and omega-3 oil, say." [79]

1.11.2.3 Flexible Electronics

"The novel device operates on magnetoresistive random access memory (MRAM), which uses a magnesium oxide (MgO)-based magnetic tunnel junction (MTJ) to store data. MRAM outperforms conventional random access memory (RAM) computer chips in many aspects, including the ability to retain data after a power supply is cut off, high processing speed, and low power consumption." [80]

1.11.2.4 Self-assembling Microchips

"Microchips, for example, use meticulously patterned templates to produce the nano-scale structures that process and store information. Through self-assembly, however, these structures can spontaneously form without that exhaustive preliminary pattern-ing. And now, self-assembly can generate multiple distinct patterns, greatly increasing the complexity of nanostructures that can be formed in a single step." [81]

1.11.2.5 Flexible, Transparent Cell Phones and Solar Cells, and Ultrafast Computers and Internet

In a quick, four-minute video, Dr Frank Koppens of the Institute of Photonic Sciences/ Nano-optoelectronics research group describes the structure of graphene, one of the most promising and rapidly emerging nanomaterials. He also outlines its applications, including future smart phones that are flexible and transparent; new types of circuitry for ultrafast processing in ultrafast phones, computers, and the internet; and flexible transparent solar cells that could be mounted on windows.

Using Lego bricks, he demonstrates how quantum dots, which are excellent light absorbers, might be mounted on a graphene substrate to make ultrasensitive, flexible, transparent photosensors that could be used, for example, as night vision cameras for cars. Finally, he describes a new generation of computers in which light runs through very tiny circuits that can be manipulated at the nanoscale. [82]

1.11.2.6 More Food, Less Energy, Less Water

This article describes how zinc nanoparticles can activate enzymes in plants without the need to use inefficient fertilizers, which are applied by spraying or in irrigation.

"When we made these enzymes more active, the plants took up nearly 11 % more phosphorus than was naturally present in the soil, without receiving any conventional phosphorous fertilization. The plants that we treated with zinc nanoparticles increased their biomass (growth) by 27 % and produced 6 % more beans than plants that we grew using typical farm practices but no fertilizer." The authors also pointed out that this "nanofertilizer also has the potential to increase plants' nutritional value." [83]

1.11.2.7 Personalized Medicine

"Cancer and neurodegenerative diseases like Alzheimer's and Parkinson's are all able to be better treated if detected early. Unfortunately, this is not always the case as symptoms may not appear until these diseases are well established. To help counteract this problem, scientists at the National Nanotechnology Laboratory (LNNano) in Brazil have created a biosensor capable of rapidly detecting molecules specifically linked to various cancers and neurological diseases.

Essentially a nanometer-size, single-layer organic transistor mounted on a glass slide, the new biosensor contains a reduced form of a peptide (a short chain amino acids; also referred to as 'small proteins') known as glutathione (GSH). This substance, when exposed to the enzyme glutathione S-transferase (GST) – associated with Parkinson's, Alzheimer's, breast cancer and a number of other diseases – creates a reaction that is detected by the transistor." [84]

1.11.3 Energy Out of Thin Air

"A team of engineers at the University of Massachusetts Amherst has recently shown that nearly any material can be turned into a device that continuously harvests electricity from humidity in the air. The secret lies in being able to pepper the material with nanopores less than 100 nanometers in diameter. The research appeared in the journal *Advanced Materials*.

"What we realized after making the Geobacter discovery," says Yao, "is that the ability to generate electricity from the air—what we then called the 'Air-gen effect'—turns out to be generic: literally any kind of material can harvest electricity from air, as long as it has a certain property."

"That property? "It needs to have holes smaller than 100 nanometers (nm), or less than a thousandth of the width of a human hair."

"This is because of a parameter known as the "mean free path," the distance a single molecule of a substance, in this case water in the air, travels before it bumps into another single molecule of the same substance. When water molecules are suspended in the air, their mean free path is about 100 nm.

"Yao and his colleagues realized that they could design an electricity harvester based around this number. This harvester would be made from a thin layer of material filled with nanopores smaller than 100 nm that would let water molecules pass from the upper to the lower part of the material. But because each pore is so small, the water molecules would easily bump into the pore's edge as they pass through the thin layer. This means that the upper part of the layer would be bombarded with many more charge-carrying water molecules than the lower part, creating a charge imbalance, like that in a cloud, as the upper part increased its charge relative to the lower part. This would effactually create a battery—one that runs as long as there is any humidity in the air." [85]

1.11.4 Solar Cells

No vision of the future of nanotech would be complete without a discussion of solar cells. This sector of "Clean Energy" is about to undergo a major revolution, with new architectures and new materials emerging. StatNano's Insight article, "Technology of the Future: Nano Insights into Solar Cells and Photovoltaics" is well worth reading, even for the average citizen [86]. Here is the abstract from that article:

"Over the past decades, nanotechnology has played a pivotal role in improving the performance of first-generation silicon solar cells and laid the foundations for the new generations of solar cells based on thin films and nanostructures. Thus far, a wide variety of nanostructured solar cells have been introduced, but many scientists from around the world are still working to develop more and more efficient solar cells; perovskite, dye-sensitized, quantum dot, polymer, copper indium gallium selenide (CIGS), copper zinc tin sulfide (SZTS), cadmium telluride (CdTe), and gallium arsenide (GaAs) solar cells are cases in point. As for silicon solar cells, the main application of nanotechnology is to enable the development of antistatic, self-cleaning, and water-repellant thin films based on polyvinylidene fluoride or organosilanes to help improve their efficiency by reducing surface contamination. Since the major advances of solar technology have, directly or indirectly, their roots in nanotechnology, it appears that the future of solar power generation is, indeed, in the hands of nanotechnology."

1.12 Themes for Thought

a. From where you are now, describe two different pathways for you to support nano-safety.
b. Nanotechnology: is it really a science of its own or is it merely an extension of chemistry, physics, and biology? State your position and defend it.
c. Download a research report on a specific sector of the Nanotechnology market and evaluate the quality of the data. Determine whether it is citing funding or revenue numbers, which sectors it is really including, and whether or not the data are a true reflection of a specific Nanotechnology or are based on nano-enabled technologies and nanointermediates.
d. Define a project you think should be commercialized, and map out the safety hurdles you perceive in moving from concept to full commercialization.
e. Describe how technology transfer from your local school, college, or university, can contribute to commercialization of a Nanotechnology project.
f. Discuss what you perceive as the greatest hurdle to an attitude of safety in nano-technology.

Bibliography

[1] Nanotechnologies Industry Association. https://nanotechia.org/about-nia-old (accessed in 2016).
[2] The NanoDataBase. Inventory for products containing nanomaterials.
[3] Feynman R. There's plenty of room at the bottom. Am Phys Soc, 1959, Cal Tech, Pasadena, CA, December 29, 1959, Caltech Engineering and Science, 23, 5, February 1960, pp. 22–36. It has been made available on the web courtesy of Zyvex, at http://www.zyvex.com/nanotech/feynman.html (accessed in 2016).
[4] The Silicon Engine. Photolithography techniques are used to make silicon devices, 1955. http://www.computerhistory.org/siliconengine/photolithography-techniques-are-used-to-make-silicon-devices/ (accessed in 2016).
[5] Drexler E. Engines of creation: the coming era of nanotechnology, Anchor Library of Science, 1986, ISBN 0-385-19973-2 (pbk).
[6] Feynman R. Tiny machines. Esalen Institute, Big Sur, CA, October 25, 1984. https://www.youtube.com/watch?v=4eRCygdW--c (accessed in 2016).
[7] Graphene: The next big (but thin) thing. https://youtu.be/Mcg9_ML2mXY, posted July 14, 2014 (accessed in 2016).
[8] Nixon A. YouTube Interview Nov 2022.
[9] Matson J. Graphene researchers Geim and Novoselov win Nobel Prize in Physics [Updated]. Scientific American, 6 Oct, 2010.
[10] Boyd J. Nanobrushes: A new carbon material to boost batteries and capacitors. The Spectrum (IEEE), 2016, Posted 6 Jul 2016. http://spectrum.ieee.org/nanoclast/semiconductors/nanotechnology/nanobrushes-a-new-carbon-material-to-boost-batteries-and-capacitors (accessed in 2016).
[11] Deng S et al. Confined, "Oriented, and electrically anisotropic graphene wrinkles on bacteria". ACS Nano, 2016, DOI: 10.1021/acsnano.6b03214. Publication date (Web): July 08, 2016. http://pubs.acs.org/doi/abs/10.1021/acsnano.6b03214 (accessed in 2016). A summary of this research was also released July 13, 2016 at Microscopy & Analysis: http://www.microscopy-analysis.com/editorials/editorial-listings/germs-add-ripples-graphene?elq_mid=10831&elq_cid=1636070 (accessed in 2016).

[12] Carbon nanotubes. History and development of carbon nanotubes (buckytubes). AZoNano, 17 Jun 2004. http://www.azonano.com/article.aspx?ArticleID=982 (accessed in 2016).

[13] Hinrichsen E. How were CNTs discovered, Apr 29, 2011. http://www.brighthubengineering.com/manufacturing-technology/115961-who-discovered-nanotubes/ (accessed in 2016).

[14] Montioux M, Kuznetsov V. Who should be given the credit for the discovery of carbon nanotubes? Carbon, 2006, 44, 1621 (ELSEVIER).

[15] Nisha CK, Mahajan Y. Birth and early history of carbon nanotubes. Nanowerk Spotlight, June 3, 2016. http://www.nanowerk.com/spotlight/spotid=43558_1.php (accessed in 2016).

[16] Радушкевич ЛВ. Структуре Углерода, Образующегося При Термическом Разложении Окиси Углерода На Железном Контакте (PDF). Журнал Физической Химии (in Russian). 1952, 26, 88–95 (Wikipedia, Carbon Nanotubes – Note CVD-86).

[17] Iijima S. Helical microtubules of graphitic carbon. Nature, 1991, 354, 56–58, DOI: 10.1038/354056a0 (Wikipedia, "Sumio Iijima", Note 1). See also Iijima's address on receiving the 2007 Balzan award: http://www.balzan.org/en/prizewinners/sumio-iijima/the-discovery-of-carbon-nanotubes-iijima (accessed in 2016).

[18] Boysen E, Muir NC. What is a quantum dot, *Nanotechnology for Dummies*, 2nd edn. http://www.dummies.com/how-to/content/what-is-a-quantum-dot.html (accessed in 2016).

[19] Wikipedia discussion of Quantum dots. https://en.wikipedia.org/wiki/Quantum_dot#cite_note-5 (accessed in 2016).

[20] Ramírez HY, Flórez J, Camacho AS. Efficient control of coulomb enhanced second harmonic generation from excitonic transitions in quantum dot ensembles. Physical Chemistry Chemical Physics, 2015, 17(37), 23938.

[21] Smith DM et al. Applications of nanotechnology for immunology. Nature Reviews. Immunology, 2013, 13, 592–605. DOI: 10.1038/nri3488. Published online July 25, 2013. http://www.nature.com/nri/journal/v13/n8/full/nri3488.html (accessed in 2016).

[22] NanoComposix. Transmission electron microscopy analysis of nanoparticles, Sept 2012 V1.1. http://50.87.149.212/sites/default/files/nanoComposix%20Guidelines%20for%20TEM%20Analysis.pdf (accessed in 2016).

[23] Foster B. TERS – Ready or Not? Am Lab, Sept 2016. http://www.americanlaboratory.com/914-Application-Notes/190745-TERS-Ready-or-Not/ (accessed in 2016).

[24] Foster B. Superresolution: Reality or a "STORM" in a Teacup. BioPhotonics, Jan 2011, Feature Article.

[25] Gogotsi Y, Huang Q. MXenes: Two-dimensional building blocks for future materials and devices. ACS Nano, 2021, 15(4), 5775–5780. Publication date: April 27, 2021. https://doi.org/10.1021/acsnano.1c03161 (downloaded 2023).

[26] Trybula W. More nanomaterials, or is it metamaterials, or semiconductors? Nano-blog.com, February 2023, http://www.nano-blog.com/?p=551 (accessed 2023).

[27] Image cited in nano-blog was taken from Babak Anasori B, Xie Y, Beidaghi M, Lu J, Hosler BC, Hultman L, Kent PRC, Gogotsi Y, Barsoum MW. Two-Dimensional, Ordered, Double Transition Metals Carbides (MXenes), https://pubs.acs.org/doi/full/10.1021/acsnano.5b03591 (downloaded 2023).

[28] Read more about Nanomaterials at https://statnano.com/nanomaterials#ixzz86LMfj2jo.

[29] Read more about which materials are getting the most new patents: https://statnano.com/nanomaterials#ixzz865FFxaTm.

[30] NNI. Big things from a tiny world. http://www.nano.gov/node/240 (accessed in 2016).

[31] NNI. Nanotechnology and energy: Powerful things from a tiny world.

[32] Learn more about NNI at nano.gov.

[33] NNI Supplement to the President's 2023 Budget. https://www.nano.gov/sites/default/files/pub_resource/NNI-FY23-Budget-Supplement.pdf (accessed in 2023).

[34] NNI FY2024 Budget. https://nsf-gov-resources.nsf.gov/2023-03/59_fy2024.pdf?VersionId=7x.iz0tdIogrA_zNfeXnXkqeLze6JAVC#:~:text=FY%202024%20Funding&text=NNI%20enables%20increased%20interdisciplinarity%20in,in%20NSE%20in%20FY%202024 (downloaded 2023).

[35] STATNano compendium of policy reports. https://statnano.com/policydocuments.

[36] Salamanca-Buentello F, Persad DL, Court EB, Martin DK, Daar AS, Singer PA. Nanotechnology and the developing world. PLoS Medicine, 2005, 2(5), e97, DOI: 10.1371/journal.pmed.0020097, Published: May 12, 2005.

[37] UN's Sustainable Development Goals. https://www.un.org/sustainabledevelopment/sustainable-development-goals/ (downloaded, 2023).

[38] Nanotechnology for a Sustainable Future: Addressing Global Challenges with the International Network4Sustainable Nanotechnology. https://pubs.acs.org/doi/10.1021/acsnano.1c10919.

[39] Grabowski R. Who is going to buy the darn thing?. Proceedings of the IEEE Electro International, June 21. 1995, pp. 69–97. Download from http://marketingvp.com/download/whois.pdf.

[40] Berger M. Debunking the trillion dollar nanotechnology market size hype. Nanowerk Spotlight, April 19, 2007. http://www.nanowerk.com/spotlight/spotid=1792.php (accessed in 2016).

[41] Cientifica. The nanotechnology opportunity report, 2nd edn., June 2003.

[42] Harper T. The year of the trillion dollar nanotechnology market? AzoNano, 2015, 2. http://www.azonano.com/article.aspx?ArticleID=3946 (accessed in 2016).

[43] Roco M. NSF nanoscale science and engineering at 20 years of NNI. https://www.nseresearch.org/2020/presentations/NNI_2020-1201_MRoco_NT%20at%20NSF%20at%2020yrs%20NNI_web.pdf (downloaded 2023).

[44] StatNano. Proliferation of Nanotechnology on WoS from 20 through June 2023. https://statnano.com/report/s29.

[45] Universities around the world offering nanotechnology programs. Statnano.com https://statnano.com/orgs (downloaded 2023).

[46] StatNano – USPTO Nanotechnology patents. https://statnano.com/news/71177/Nanotechnology-Published-Patent-Applications-in-USPTO-Number-and-Annual-Growth-Rate-during-the-Past-20-Years (downloaded 2023).

[47] Nanotechnology patents awarded 2020–2023. https://statnano.com/news/71177/Nanotechnology-Published-Patent-Applications-in-USPTO-Number-and-Annual-Growth-Rate-during-the-Past-20-Years#ixzz8656UPwuX (downloaded 2023).

[48] RDWorldMagazine – R&D Global Funding Forecast. Download: https://forecast.rdworldonline.com/. Copy received with compliments of their editor.

[49] Vance M et al. Nanotechnology in the real world: Redeveloping the nanomaterial consumer products inventory. Beilstein Journal of Nanotechnology, 2015, 6, 1769–1780.

[50] Bowman DM, Hodge GA. Nanotechnology: Mapping the wild regulatory frontier, https://doi.org/10.1016/j.futures.2006.02.017 (downloaded 2023).

[51] Lux Report. Nanotechnology update: Corporations up their spending as revenues for nano-enabled products increase, February 17, 2014. https://portal.luxresearchinc.com/research/report_excerpt/16215 (accessed in 2016).

[52] Roco M. NSF, Market report on emerging nanotechnology, NSF Media Advisory 14–004. https://www.nsf.gov/news/news_summ.jsp?cntn_id=130586 (accessed in 2016).

[53] BCC Research. Nanotechnology: A realistic market assessment, NAN031F, Nov 2014. http://www.bccresearch.com/market-research/nanotechnology/nanotechnology-market-assessment-report-nan031f.html (accessed in 2016).

[54] BCC Report: Global Nanotechnology Market. https://www.bccresearch.com/market-research/nanotechnology (downloaded 2023).

[55] NBIC+: StatNano News site for nanotech, biology, IT, computation, etc. https://statnano.com/news.

[56] Ghaffarzadeh K. Graphene, 2D materials and carbon nanotubes: Markets, technologies and opportunities 2016–2026, IDTechEx. http://www.idtechex.com/research/reports/graphene-2d-materials-and-carbon-nanotubes-markets-technologies-and-opportunities-2016-2026-000465.asp (accessed in 2016).

[57] Why hasn't graphene taken over the world… yet? https://www.youtube.com/watch?v=IybSD1QlPSk.

[58] GAC press release. https://www.gac-motor.com/en/media/newsdetail/id/166.html#:~:text= Recently%2C%20GAC%20Group%20announced%20a,phase%20of%20actual%20vehicle%20testing (downloaded 2023).

[59] Universal Matter, Our Solution. https://www.universalmatter.com/our-solution/ (downloaded 2023).

[60] Transparency Market Research. Nanotechnology drug delivery market to touch US$ 11.9 billion in 2023, Pharmacy News Today, April 4, 2016. http://www.pharmacynewstoday.com/nanotechnology-drug-delivery-market-to-touch-us-119-billion-in-2023-transparency-market-research (accessed in 2016).

[61] Precedence Research. (n. d.). Nanotechnology Drug Delivery Market (By Technology: Nanocrystals, Nanoparticles, Liposomes, Micelles, Nanotubes, Others; By Application: Neurology, Oncology, Cardiovascular/Physiology, Anti-Inflammatory / Immunology, Anti-infective, Others) – Global Industry Analysis, Size, Share, Growth, Trends, Regional Outlook, and Forecast 2023–2032. https: //www.precedenceresearch.com/nanotechnology-drug-delivery-market#:~:text=The%20global% 20nanotechnology%20drug%20delivery,8.13%25%20between%202023%20and%202032.

[62] TechnavioPlus, Food Nanotechnology Market by Application and Geography – Forecast and Analysis 2021-2025, Published: Jan 2022. https://www.technavio.com/talk-to-us?report=IRTNTR71940&type= sample&src=report (downloaded 2023).

[63] Cientifica. Wearables, smart textiles and nanotechnology: applications, technologies and markets, May 2016. http://www.cientifica.com/ (accessed in 2016).

[64] Research and Markets (2023, April 28). Smart Textiles Global Market is Projected to Reach $13.8 Billion by 2030: Smart Garments Emerge as Sub-Sets of Wearable Technology. GlobeNewswire News Room. https://www.globenewswire.com/en/news-release/2023/04/28/2657404/28124/en/Smart-Textiles-Global-Market-is-Projected-to-Reach-13-8-Billion-by-2030-Smart-Garments-Emerge-as-Sub-Sets-of-Wearable-Technology.html#:~:text=The%20global%20market%20for%20Smart,the%20analysis% 20period%202022%2D2030.

[65] MarketsandMarkets. Internet of nano things market by communication type (short & long distance communication), by nano components & devices (cameras, phones, scalar sensors, processors, memory cards, power systems, antennas & transceivers): Worldwide forecast & analysis (2016–2020), September 2014. http://www.marketsandmarkets.com/Market-Reports/internet-nano-things-market-10414659.html (accessed in 2016).

[66] Prudour Private Limited (2023, March 24). Internet of Nano Things Market is Estimated to Showcase Significant Growth of USD 123.5 Bn in 2021 With a CAGR 22.38 %. Enterprise Apps Today. https: //www.enterpriseappstoday.com/news/internet-of-nano-things-market-is-estimated-to-showcase-significant-growth-of-usd-123-5-bn-in-2032-with-a-cagr-22-38.html.

[67] Highlights of Recent Research on the Environmental, Health, and Safety Implications of Engineered Nanomaterials, September 01, 2017. https://www.nano.gov/Highlights-Federal-NanoEHS-Report (downloaded 2023).

[68] It has been the Author's experience that an EHS officer is assigned from initial stages in all early stage nanotech companies that she has encountered.

[69] Applying a lab safety culture to nanotechnology: Educating the next generation. http://www.nano. gov/node/1620, April 29, 2016.

[70] NEHI: Description and goals. https://www.nano.gov/about-nni/working-groups/nehi (downloaded 2023).

[71] Nanolink Presentations. http://www.nano-link.org/component/docman/cat_view/.

[72] MetPhast Nanotechnology & Safety Course. http://www.metphast.umn.edu/services/ nanotechnology-health-safety/ (downloaded 2023).

[73] International Association of Nanotechnology, Nanossafety training courses: https://www.ianano.org/ Site/Training/index.html (downloaded 2023).

[74] Oxford University Courses in Nanontechnology and Nanomedicine. https://www.conted.ox.ac.uk/ about/nanotechnology-and-nanomedicine (downloaded 2023).

[75] United Nations Institute for Training and Research (Unitar) Nanomaterials Safety Course. https: //event.unitar.org/full-catalog/nanomaterials-safety-course. More general info: https://event.unitar. org/.

[76] NanoDays, sponsored by NISE. https://www.nisenet.org/nanodays (downloaded 2023).

[77] National Nanotechnology Initiative Strategic Plan (October 2021). https://www.nano.gov/sites/ default/files/pub_resource/NNI-2021-Strategic-Plan.pdf (downloaded 2023).

[78] Perkle JM. DNA origami robots, biotechniques, 05/06/2014. http://www.biotechniques.com/news/ DNA-Origami-Robots/biotechniques-351627.html?autnID=331944#.V45zL_krLIU (accessed in 2016).

[79] Are you ready for your nanotechnology engineered wine? Nanowerk, February 12, 2007. http: //www.nanowerk.com/news/newsid=1441.php (accessed in 2016).

[80] Scientists develop plastic flexible magnetic memory device, Nanowerk News, Jul 19, 2016. http: //www.nanowerk.com/nanotechnology-news/newsid=43976.php (accessed in 2016).

[81] Smarter self-assembly opens new pathways for nanotechnology. R&D Magazine, Mon, 08/08/2016 – 12:30 pm. https://www.rdmag.com/news/2016/08/smarter-self-assembly-opens-new-pathways-nanotechnology (accessed in 2016). https://www.youtube.com/embed/GlQZpws7hd4 (accessed in 2016).

[82] Smarter self-assembly opens new pathways for nanotechnology. R&D Magazine, Mon, 08/08/2016 – 12:30 pm. https://www.rdmag.com/news/2016/08/smarter-self-assembly-opens-new-pathways-nanotechnology (accessed in 2016). https://www.youtube.com/embed/GlQZpws7hd4 (accessed in 2016).

[83] Raliya R. How nanotechnology can help us grow more food using less energy and water. Nanowerk News, May 26, 2016. http://www.nanowerk.com/nanotechnology-news/newsid=43511.php (accessed in 2016).

[84] Jeffrey C. Sensor detects signs of cancer, Alzheimer's, and Parkinson's, May 24, 2016. http://www. gizmag.com/cancer-alzheimers-parkinsons-biosensor-transistor/43475/ (accessed in 2016).

[85] Engineers at UMass Amherst harvest abundant clean energy from thin air, 24/7. https://www.umass. edu/news/article/engineers-umass-amherst-harvest-abundant-clean-energy-thin-air-247.

[86] StatNano Insight article: "Technology of the Future: Nano Insights into Solar Cells and Photovoltaics." https://statnano.com/en/page/5140.

Eylem Asmatulu

2 The World of Engineering Nanomaterials

2.1 Introduction

2.1.1 How Did Engineering Nanomaterials Evolve?

The first introduction to nanotechnology began with American physicist Richard Feynman stating in a speech at an American Physical Society meeting at Caltech in 1959, "There is plenty of room at the bottom" [1]. Feynman specified that new processes could be developed at smaller scales for the manipulation of atoms and molecules, using new tools to investigate the behavior of materials at those smaller scales. Scaling changes, the magnitude of many physical phenomena, such as gravity (which become less significant) and surface tension and van der Waals attraction (which become more significant) [2].

A scientist at Tokyo University of Science, Norio Taniguchi, was the first to use the term "nanotechnology" in a 1974 conference. He described an example of semiconductor processes, such as ion beam milling and thin-film coating, showing characteristic control at the nanoscale [1]. He described it this way: "'Nano-technology' mainly consists of the processing of separation, consolidation, and deformation of materials by one atom or one molecule." In 1981, the American engineer Eric Drexler published his first paper on nanotechnology. Furthermore, Drexler was credited with the development of molecular nanotechnology, leading to manufacture of nanosystems machinery. In the 1980s, Zurich scientists at IBM invented scanning tunneling microscopy (STM). This invention was followed by the invention of atomic force microscopy (AFM), which allowed scientists to see materials for the first time at near the atomic level. Figure 2.1 shows an example of images at the nanoscale, obtained using AFM and STM.

In the 1980s, the accessibility of supercomputers enhanced many activities, including modeling and simulation, atomic scale visualization and characterization, and experimental synthesis activities, which in turn encouraged nanoscale research activities. Another innovation came in 1985 with the discovery of fullerenes (buckyballs), which are perfectly spherical and have 60 carbon atoms. A significant achievement was made in 1990 when a team of physicists wrote the word "IBM" using 35 xenon atoms. The earlier discovery of buckyballs pioneered new discoveries, such as carbon nanotubes (CNTs) in 1991. Applications involving CNTs are growing because of their extraordinary thermal, mechanical, and electrical properties. CNT characteristics that make them promising in nanotechnology are that they are 100 times stronger than steel and six times lighter in weight. Along the same lines, new studies on semiconductors and nanocatalysts have led to many new developments for quantum dots (QDs), which have properties between those of bulk semiconductors and discrete molecules. All of these innovations encouraged industrialized nations to form new nanotechnology initiatives in

https://doi.org/10.1515/9783110781830-002

ZT042 F1301 Topography, ZT042F_HDF

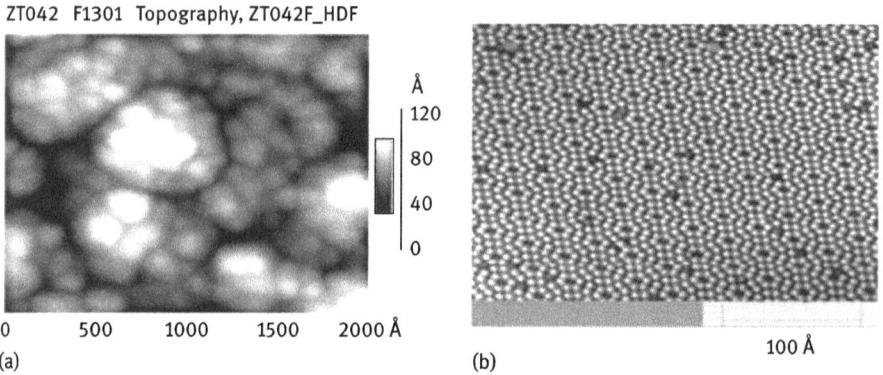

(a)

(b)

100 Å

Figure 2.1: Images taken at the nanoscale: (a) AFM image of $ZnSnO_4$/glass showing grains smaller than 100 Å, and (b) STM image of Si (111) 7 × 7 reconstruction. Source: National Renewable Energy Laboratory (NREL), 2017 [3].

the early 2000s, which in turn has led to a global increase in nanotechnology activities. Working with both academia and industry, the US Interagency Working Group on Nanotechnology (IWGN), established by the Office of Science and Technology Policy (OSTP), created the US National Nanotechnology Initiative (NNI) to support more advanced studies in this field [1].

Today, the major focus of research around the world is to study nanoscale properties, synthesis of different materials, characterization techniques, safety concerns, and applications to make useful devices and processes, as well as gain economic benefit from these unique materials. Although nanomaterials offer limitless possibilities, they carry new challenges for detecting and handling potential safety and health hazards to scientists, engineers, students, and consumers. Based on their core components, nanomaterials can take the form of nanotubes, nanoparticles, nanowires, nanofibers, nanowhiskers, nanofilms, and nanocomposites. Even well-known bulk materials can react differently both health-wise and in the environment when they are nanosized. Because the variety of nanomaterials is enormous, there are no exact rules and regulations for many of them [1–4]. There is growing awareness of the importance of educating future engineers and scientists about this developing field, in addition to addressing the safety and health aspects [1]. This chapter provides up-to-date information on engineering controls and safe work practices to be followed when working with nanomaterials in research and teaching laboratories [4].

2.2 Stabilization of Nanomaterial Shape

2.2.1 Surfactants

Most nanoparticles are insoluble in water and have high binding affinities to each another, so stabilizing agents are necessary to reduce agglomeration and increase aqueous dispersion of nanoparticles. Many cationic, anionic, nonionic, and amphoteric surfactants have been used for stabilizing nanomaterials in the aqueous phase. However, some of these surfactants have strong cytotoxicity effects on health and the environment. For example, cetyl trimethylammonium bromide (CTAB) has been used for synthesizing and stabilizing gold nanorods, but has a strong detergent effect that seriously increases nanorod cytotoxicity in many biological applications. Use of sodium dodecyl sulfate (SDS) and sodium dodecylbenzenesulfonates (SDBS) surfactants for CNT dispersion also causes significant toxic effects compared with other alternatives. Thus, surfactant selection for dispersion of nanoparticles must be considered cautiously, and proper research should be performed to compare surfactant toxicity levels for different nanomaterials [5]. Some polymeric materials can be used to stabilize nanomaterials through a steric stabilization process. Manipulating surface changes and adding coagulants can significantly stabilize nanomaterials in liquid suspensions.

2.2.2 Nanomaterial Shape and Stabilization

2.2.2.1 Stabilization of CNTs

CNTs can be produced using various chemical and physical methods [6]. Aggregation is a challenge for most nanomaterial applications and weakens the distinct properties of the individual CNTs. As a result of strong van der Waals attraction, single-wall carbon nanotubes (SWCNTs) pack into crystalline ropes that aggregate into tangled networks, which are very difficult to separate under normal conditions. A simple procedure described for dispersing as-produced nanotube powder in aqueous solutions involves gum Arabic and a washing detergent. Because of physical adsorption of the polymer, a stable dispersion of full-length, well-separated, individual nanotubes can be formed in a single phase [7]. Nitric acid and sulfuric acid stabilizations can also be used for the same purpose.

2.2.2.2 Stabilization of Nanofibers

Nanofibers with an average diameter average 10–500 nm can be produced using various polymeric precursors (e. g., polyacrylonitrile, polyvinylchloride, and polystyrene) via an electrospinning technique. Some nanofibers have been stabilized at temperatures of

250–280 °C for 1–3 h, followed by carbonization at 1000 °C, resulting in the fabrication of carbon nanofibers (CNFs). Characterization techniques to obtain information about the morphology, thermal properties, and chemical structures of nanofibers include scanning electron microscopy (SEM), transmission electron microscopy (TEM), differential scanning calorimetry (DSC), Fourier transform infrared spectroscopy (FTIR), glass transition temperature, X-ray diffraction (XRD), and X-ray photoelectron spectroscopy (XPS) [8]. To stabilize individual nanofibers in water, similar techniques can be applied with CNT procedures.

2.2.2.3 Stabilization of Nanowhiskers

A number of different nanowhiskers have been developed and characterized for different industrial applications. Cellulosic nanowhiskers can be sterically prepared using hydrochloric acid hydrolysis of cotton powders and subsequent surface grafting of monomethoxy poly(ethylene glycol) (mPEG) [9]. Based on the structures and elemental distributions of nanowhiskers, steric techniques can be utilized to stabilize the nanowhiskers for a number of days, or even weeks. The dispersion process allows nanowhiskers to mix well with the matrix materials and create high-performance, mechanically robust nanostructured materials for various industrial applications, such as aircraft, manufacturing, energy conversion, and electronics [6, 10].

2.2.2.4 Stabilization of Nanorods

Nanorods can be produced using a wet chemical precipitation process and stabilized using a variety of surfactants and coagulants. Stable aqueous dispersions of citrate-stabilized gold nanorods (Cit-GNRs) have been made by surfactant exchange from CTAB-stabilized GNRs, utilizing polystyrenesulfonate (PSS) as a detergent. To monitor the surfactant exchange process, FTIR, surface-enhanced Raman scattering (SERS), and XPS techniques are usually chosen. Cit-GNRs are perfectly stable at low ionic strength and are relatively conducive to further ligand exchange without any loss of dispersion stability in a solution [11]. These stable nanorods can be employed in design and manufacture of bio- and nanosensors, nanocomposite layers, electric probes, and agents for targeted drug delivery [6].

2.2.2.5 Stabilization of Nanospheres

Nanospheres are another set of newly developed nanostructured materials and are made in different ceramic and polymeric forms, such as silica nanospheres. The difficulty with silica nanospheres is aggregation, which often occurs at low pH and at neutral and high pHs in a high-salt medium. Modifying the silicate surface with trihydroxysilyl propionic acid can overcome this issue. In experimental studies, 5 g of 3-(triethoxy silyl) propionitrile and 4.2 g of KOH were mixed and heated to a boiling temperature, while stirring on a hot plate. The heat was then adjusted to maintain a steady reflux of water in the condenser tube. At the end of 24 h, refluxing ammonia gas was released. Using a rotary evaporator at below 60 °C, the carboxylic acid derivate was saved, and excess water and ether were removed. The resulting transparent solution was diluted to 10 mL and analyzed by FTIR. It was concluded that 2.4 M 3-trihydroxysily propionic acid and the modified nanospheres were significantly stabilized against aggregation, even under physiological salt concentrations [12].

2.2.2.6 Stabilization of Magnetic Nanoparticles

Most magnetic nanoparticles (MNPs) correspond to iron-based nanoparticles (e. g., superparamagnetic iron oxides [SPIONs], zero-valent iron, core–shell Fe/Au, or Fe_xO_y/Au nanoparticles and ferrites); a smaller number correspond to magnesium oxide, nickel, and cobalt nanoparticles. Magnetic nanoparticles can be stabilized in nonaqueous solvents or water for short or long periods of time. MNPs are usually dispersed through functionalization by water-soluble compounds (e. g., sulfonic acid disodium salts, porphyrins, soluble polymers, citric acids, tetramethylammonium hydroxide, and calixarenes) after fabrication using chemical coprecipitation, thermal decomposition, microwave heating, and ultrasonication techniques. To increase MNP dispersion rates, polyols are frequently used. MNP stabilization in solutions can be achieved with the help of monomeric, inorganic, and polymeric compounds [13].

2.2.2.7 Stabilization of Nanoflakes

Nanoflakes, such as phosphorene, graphene, and boron nitrate, have been produced and functionalized for various industrial and technological purposes. Phosphorene has been gaining considerable attention because of its robust direct-band gap and high-charge mobility features in semiconductor applications. Phosphorene is a two-dimensional (2-D) semiconductor and allotrope of phosphorus. The first time that phosphorene was isolated by using a mechanical exfoliation process was in 2014. Today, phosphorene can be produced by mechanical or liquid exfoliation; however, direct epitaxial phosphorene growth is a challenge that significantly delays mass production and application of phosphorene nanoflakes. The stability of a nanoscale cluster or flake on a substrate is crucial in epitaxial growth. The stability of phosphorene nanoflakes is strongly reliant on the strength of interaction between the substrate and the nanoflake. For instance, strong interaction (0.75 eV/P atom) with the Cu(III) substrate breaks down the phosphorene nanoflakes, whereas weak interaction (0.063 eV/P atom) with hexagonal boron nitride (h-BN) substrate fails to stabilize its 2-D structure. Substrates with a moderate interaction (about 0.35 eV/P atom) could stabilize the 2-D structures of the nanoflake on a realistic time scale [14, 15].

2.3 Classification and Labeling of Nanomaterials

2.3.1 What Are Nanomaterials?

Nanosized particles can be found in nature. They include volcanic ash, automobile exhaust gas, cosmic dust, fire dust, windblown fine particles, and ocean mist. They can also be created from a variety of sources, such as metallic, polymeric, semiconductor, ceramic, and composite particulates, which are all man-made. To be considered a nanomaterial, one dimension of the material should be less than 100 nm, to which is attributed exclusive chemical, biological, physical, and physicochemical properties over bulk particles. The smallest size that the human eye can see is about 0.01 mm (~10 μm), so nanosized particles can neither be seen with the naked eye nor with a conventional microscope; instead, SEM, STM, TEM, and AFM techniques must be employed [16, 17]. Figure 2.2 shows a comparison of microscopic organisms and devices at the nanoscale [17].

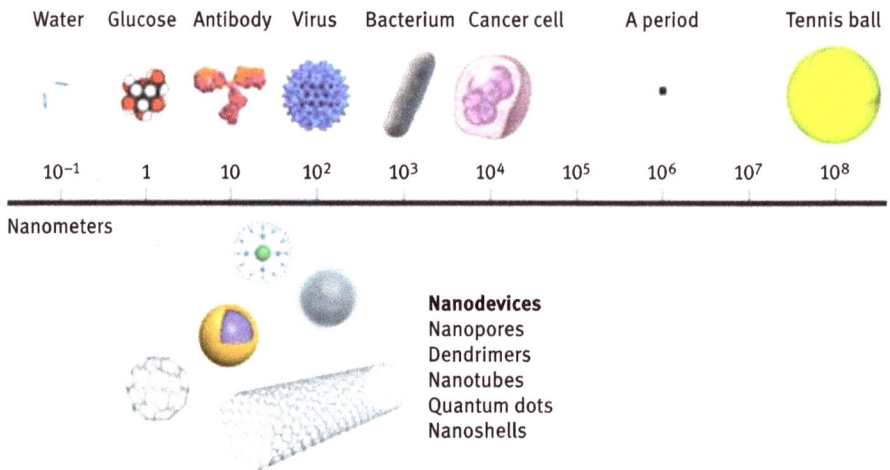

| Water | Glucose | Antibody | Virus | Bacterium | Cancer cell | | A period | | Tennis ball |

| 10^{-1} | 1 | 10 | 10^2 | 10^3 | 10^4 | 10^5 | 10^6 | 10^7 | 10^8 |

Nanometers

Nanodevices
Nanopores
Dendrimers
Nanotubes
Quantum dots
Nanoshells

Figure 2.2: Comparison of microscopic organisms and devices at the nanoscale [17].

2.3.2 Types of Nanoparticles

A number of different nanomaterials are used in a wide variety of applications in science, technology, defense, and medicine [18]. Nanoparticles fall into three major categories: naturally occurring, incidental, and engineered. Human beings have been generating nanoparticles accidentally or intentionally long before they could see the m using high-magnification electron microscopes.

2.3.2.1 Naturally Occurring Nanomaterials

Naturally occurring nanomaterials include volcanic ash, exhaust gas, cosmic dust, fire dust, windblown fine particles, ocean mist, mineral composites, and biological entities (e. g., viruses). The action of waves on rocks eventually reduces the rock surface. Some of the particles created by wave action are nanosized.

2.3.2.2 Incidental Nanoparticles

Incidental nanoparticles are the byproducts of human activity, have poorly controlled shapes and sizes, and may cause disease and environmental concerns. Many common daily sources contain incidental nanoparticles, such as cooking smoke, diesel exhaust, and welding fumes. Furthermore, incidental nanoparticles are emitted during activities such as sandblasting, waterjet cutting, crushing and grinding, metallurgical coke manufacturing, mining and blasting, and oil refining. Table 2.1 shows some of the incidental nanoparticles occurring from daily activities and their possible health effects [19].

Table 2.1: Possible health effects cause by incidental nanoparticles [20].

Source of incidental nanoparticles	Possible health effects
Diesel and other exhaust	Cancer and respiratory disease
Cooking smoke	Pneumonia, chronic respiratory disease, and even lung cancer
Welding fumes	Metal fume fever, infertility, benign pneumoconiosis
Industrial effluents	Asthma, atherosclerosis, chronic obstructive pulmonary disease
Waterjet cutting	Silicosis, respiratory disease
Sandblasting	Silicosis, respiratory disease
Crushers and fine grinders	Silicosis, respiratory disease

2.3.2.3 Engineered Nanoparticles

Engineered nanoparticles consist of any manufactured particles with precise nanoscale dimensions, shapes, and compositions. At least one dimension of these nanoparticles is in the range of 1–100 nm. Examples of this category include different metals and alloys, QDs, buckyballs/fullerenes, graphene, CNTs, sunscreen pigments, nanocapsules, nanofilms, nanocomposites, nanofibers, and nanowires [20]. Engineered nanoparticles can be simple or complex and are easily obtained with different chemical compositions, such as a gold core covered in a shell of silica and coated with specific antibodies and polymers [19].

2.3.2.3.1 Dimensions
It is necessary to classify nanomaterials on the basis of the number of dimensions, because their shape (morphology) plays an important role in their toxicity. Classification of nanomaterials is dependent on the number of dimensions in the nanoscale range (<100 nm). One-dimensional nanomaterials (e. g., coatings, thin films, and multilayers) have one dimension at the nanometer scale. In 2-D nanomaterials (e. g., fibers, tubes, and wires), two dimensions are at the nanoscale. Three-dimensional nanomaterials (e. g., quantum dots, hollow spheres, and box-shaped graphene) have all three dimensions <100 nm [21].

2.3.2.3.2 Morphology
Flatness, sphericity, and aspect ratio are the morphological characteristics of nanomaterials. Overall classification is into high- and low-aspect ratio particles. Figure 2.3 shows the classification of nanostructure materials according to dimensions, morphology, composition, agglomeration, and uniformity states. High-aspect-ratio nanoparticles contain nanotubes and nanowires in various shapes, such as zigzags, belts, helices, and nanowires of various diameters and lengths. Low-aspect ratio nanoparticles contain spherical, oval, cubic, helical, prism, or pillar shapes. These nanoparticles may exist as powders, suspension, or colloids.

2.3.2.3.3 Phase Compositions
Nanoparticles can be made from a single component material or a composite of several materials. Single-phase solids include crystals, amorphous particles, and layers. Multiphase solids can be matrix composites or coated particles. Multiphase systems include colloids, aerogels, and ferrofluids.

2.3.2.3.4 Nanoparticle Uniformity and Agglomeration
Nanoparticles can occur as dispersed aerosols, suspended colloids, or in an agglomerate state (Figure 2.3) as a result of their chemistry and electromagnetic properties. For in-

1) Dimensionality

a) 1D   b) 2D   c) 3D

| Thin films or surface coatings | Fixed long nanostructures | Fixed small nanostructures |
| Do not pose health risks | Thick membranes with nanopores | Membranes with nanopores |

| May pose health risks | Free long aspect ratio nanowires | Free small aspect ratio nanoparticles |

2) Morphology

a) High-aspect ratio   b) Low-aspect ratio

Nanowires	Nanospherical
Nanohelices	Nanohelices
Nanozigzags	Nanopillars
Nanopillars	Nanowires
Nanotubes	Nanopyramids
Nanobelts	Nanocubes
	Various

3) Composition

a) Single material   b) Composites

Compact	Coated
Hollow (spherical or nanotubes)	Encapsulated
	Barcode
	Mixed

4) Uniformity & agglomeration state

a) Isometric b) Inhomogeneous

| | | Dispersed |
| | | Agglome-rates |

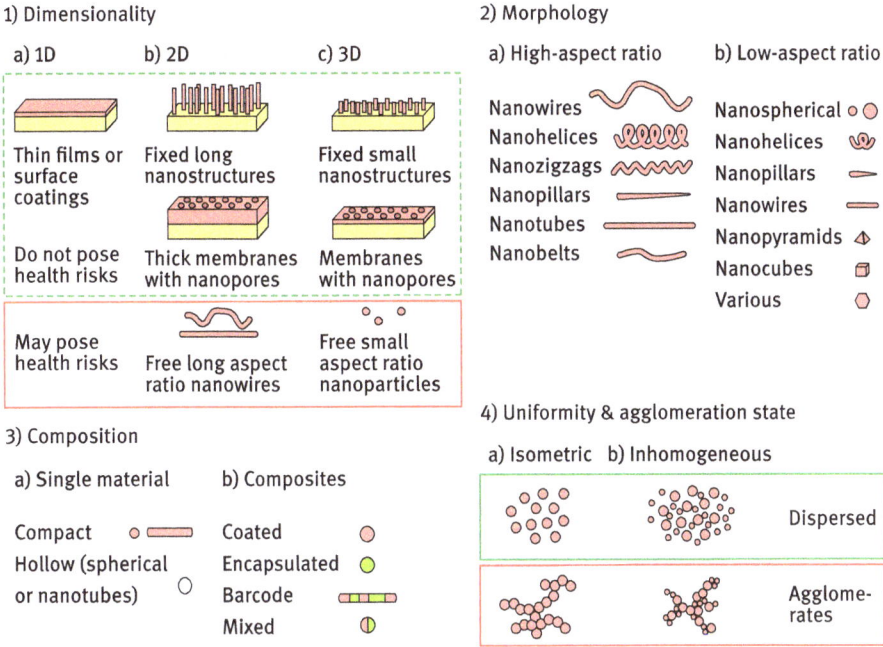

Figure 2.3: Classification of nanostructured materials on the basis of dimensions, morphology, composition, uniformity, and agglomeration state. Source: Pipal et al. (2014) [22].

stance, magnetic nanoparticles have a tendency to cluster under a magnetic field, forming an agglomerate state if their surfaces are not coated with a nonmagnetic material. Furthermore, nanoparticle agglomeration, size, and surface reactivity, along with shape and size, need to be considered when choosing health and environmental regulations for new materials [20].

2.3.3 Labeling of Nanomaterials

2.3.3.1 Nanoproducts in the Market

The Woodrow Wilson Center set up a project on emerging nanotechnologies (PEN), which is a foundation program that analyzes consumer nanoproducts. The PEN's consumer products inventory (CPI) offers a complete list of nanoproducts. As of 2014, the nanotechnology CPI contained 1814 products or product lines. Consumer nanoproducts have a wide range of applications (e. g., sporting goods, clothing, personal care products, and medicine) as well as contributions to faster and stronger cars and planes, more powerful computers and satellites, better micro- and nanochips, and longer lasting batteries [23]. Although some of the nanoproducts available on the mar ket are completely free of health concerns, the safety of remaining products is still unknown. Exposure to nanomaterials can occur during the experimentation, production, transportation, and consumption of nanoproducts. Examples of exposure to a nanoproduct include using aerosol sprays, applying sunscreen body lotion, taking medicine, or consuming supplements that contains nanomaterials.

2.3.3.2 Nanomaterial Labeling in the USA: Present Status

The USA regulates labeling on many types of consumer products, although none of the regulations require product labeling to indicate the existence of nanomaterials or the use of nanotechnologies. Even though nanomaterial labeling should be required for a variety of products, currently this requirement has only been considered for food and cosmetic products.

2.3.3.2.1 Food Labeling

The primary responsibility of the US Food and Drug Administration (FDA) is to focus on food safety in the USA, which relies on the multiple legal authorities that govern food, food ingredients, and dietary supplements. The FDA follows the Federal Food, Drug, and Cosmetic Act (FFDCA), and other laws governing the evaluation and approval of new food and color additives before they can be marketed. The FDA also focuses on postmarket regulatory tools, including labeling, to ensure product safety for other food ingredients and dietary supplements. These labels should be carefully designed with the exact details of the ingredients, and should not be misleading to consumers and agencies. To date, the FDA has not provided clear guidance for all nanomaterials used in consumer products (regarding benefits and the risk of nanomaterials) and their labeling. However, in 2007, the FDA published a Nanotechnology Task Force report on labeling, which discussed whether all food products containing nanomaterials should disclose details on their labels [24].

2.3.3.2.2 Cosmetic Labeling

The US cosmetics laws and regulations do not require nanotechnology-specific labeling of cosmetics. Similar to the situation for food, the FDA regulates cosmetic labeling, which mainly relies on FFDCA and FPLA laws. For some color additives, there are a few exceptions for premarket notification, but not for postmarket tools, including labeling and monitoring. The FFDCA mandates that agencies and manufacturers eliminate adulterated and misbranded products from the market, through legal action if necessary. If the product labeling fails to include the required information, the cosmetic product is misbranded. Cosmetic labeling should include the materials in the cosmetics and avoid false or misleading information. According to FDA regulations, labels should consist of a list of ingredients and include all relevant warnings for consumers. The manufacturers perform some safety analysis, but are not required to share their findings with the FDA.

Improperly labeled products must bear a warning label, informing consumers that the cosmetics company substantiates the safety of their ingredients and/or cosmetic products.

To date, US regulators and legislators have not specified a real need to introducing comprehensive nanomaterial labeling for many cosmetic products currently available on the market. In 2007, although the FDA considered the labeling of nanomaterials for food and cosmetic products, it was not recommended. This decision was mainly because of a lack of understanding of the risks of nanomaterials. Since 2007, many agencies and manufacturers have not considered comprehensive nanomaterial labeling on products associated with nanomaterials.

For a number of reasons, consumer labeling of nanomaterials has become a serious issue from many public perspectives. According to PEN (Woodrow Wilson Center), more than 1814 nanoproducts exist on the market. Recently, calls have been rising for mandatory consumer labeling of nanomaterials. In the USA, there is no general labeling requirement for many nanomaterials; however, some specific rules governing product labeling in the food and cosmetics area should apply to nanomaterials and nano-enabled products [24].

2.4 Toxicity of Nanomaterials

2.4.1 Particle Size

Surface area substantially increases with a reduction in the size of nanoparticles. Increased surface area allows some additional chemical interactions to take place at the nanosurface, which in turn can increase reactivity and toxicity effects. Nanosized particles can easily pass through the cell membrane and interact with DNA structures to cause damage. Compared with the same materials containing larger particles, materials containing nanosized particles (<100 nm) can cause greater adverse health effects, such as inflammation, chronic respiratory illnesses, and cancer.

2.4.2 Surface Chemistry

It has been reported that aggregated nanoparticles are less toxic than individual nanoparticles, because their relative surface area is drastically reduced. Surface chemistry determines the aggregation levels of nanomaterials in dry and wet conditions. Surface chemistry also determines the wetting properties and surface characteristics, which control some specific chemical reactions that remain active or passive for surface-controlled nanomaterial growth.

2.4.3 Surface Charges

Surface charge densities of nanomaterials predominantly affect toxicity levels, so high surface charge densities can result in higher cytotoxic effects than those with low charge densities. Particles with positive or negative high surface charges remain in suspended form for a longer period of time than those particles with low surface charge. Particles with high surface charges can create additional damage to cells and surrounding tissue, because they react more intensely with cell membranes [20]. Zeta potential is the electrostatic property that measures the colloidal stability of nanomaterial samples in liquid suspension. This is closely associated with the particle surface charge and severely influences the aggregation state. Nanoscale colloids having a low zeta potential tend to aggregate, which can be better for reducing toxicity levels. The resultant aggregation can be observed using particle size and concentration measurements due to the enlarged size of aggregates [25].

2.4.4 Surface Area

Nanoparticles have a larger surface area and higher particle number per unit mass than microparticles. Because material in nanoparticulate form offers a greater surface area for chemical reactions, reactivity with the surrounding tissue is also greater. According to Uddin (2021) [48] and Oberdörster (2001) [49], surface area plays a major role and is highly associated with particle-caused adverse health effects [26]. Oxidative stress occurs as the result of free radicals produced by nanoparticles in the human body. Inflammation, cell destruction, and genotoxicity can occur as a result of biological oxidative stress. The particle surface of free radicals can activate the redox cycle and improve particle toxicity levels. As particle size decreases, the number of atoms/ molecules on the surface increases, in turn increasing interactions with the surrounding tissues and environment. This means an increased chance of additional chemical reactivity of a particle and, thus, production of reactive oxygen species (ROS) and free radicals. ROS-induced oxidative stress is the original mechanism of nanomaterial toxicity, which can cause DNA damage, cell membrane disruption, cell leakage, and interference with cell signaling. ROS have also been determined in the secondary toxicity effects of nanomaterials, which cause oxidation of proteins and the release of hazardous components [25].

2.5 Exposure Assessment

2.5.1 Exposure Limit for Nanoparticles

Because of their unique physical, chemical, physicochemical, and biological properties, nanomaterials are appealing materials for the twenty-first century. However, the available information on possible environmental and health effects for some of these nanomaterials is limited. Therefore, the National Institute for Occupational Safety and Health (NIOSH) has provided some endorsements for limiting worker exposure to nanoparticles through standard practices, which includes respiratory protection and other safe laboratory practices. The NIOSH recommended exposure limits (RELs) for some forms of engineered nanoparticles are mainly related to their masses, but also to the special chemical and physical properties of nanoparticles, such as shape, surface energy, surface area, and reactivity [27].

The occupational exposure limit (OEL) is one of the main tools used for preventing occupational disease from specific exposure. Providing risk managers and health authorities with a quantitative health basis can be good work practice for measuring the effectiveness of nanomaterial safety procedures (use of engineering controls and other general laboratory guidance) [28]. Thus, OELs for many nanomaterials can be useful in decreasing the health risk from those materials for workers who are exposed to nanoparticles.

Several new nanomaterials are developed every year and enter the market. Currently, no regulatory standards for specific nanomaterials have been established in the USA. An REL has been recommended via NIOSH to the Occupational Safety and Health Administration (OSHA) for use as an acceptable exposure limit [29]. More recently, NIOSH has stated that the recommended exposure limit for titanium dioxide (TiO_2) is 2.4 mg/m^3 for a fine compound and 0.3 mg/m^3 for an ultrafine compound (including engineered nanoscale material), assuming time-weighted average (TWA) concentrations for up to 10 h per day during a 40-h work week (Table 2.2) [30, 31]. Another statement by NIOSH is that worker exposure should be limited to no more than 1 µg/m^3 for nanotubes and nanofibers [30, 31].

Without any governmental direction on exposure limits, some of the manufacturers producing nanomaterials have established suggested OELs for their nanoproducts. For example, the Bayer company has developed its own OEL of 0.05 mg/m^3 for Baytubes® (multiwalled CNTs) [45]. For Nanocyl CNTs, the no-effect concentration in air was estimated to be 2.5 µg/m^3 for an 8-h/day exposure, which may offer protection for workers [28, 29, 46].

Similarly, the European legislation REACH requires a manufacturer-driven derived no-effect level (DNEL) for material they bring to market (ECHA, 2010) [50, 51]. DNELs can be used to create acceptable exposure limits for workers [47]. Table 2.2 reviews the general efforts to derive a mass-based OEL or DNEL for numerous nanoparticles. Table 2.2 also demonstrates the large differences for CNTs with a "similar" identity.

Because of limited information about exposure limits for all nanomaterials and their products, employers and researchers should minimize worker exposure to nano-

Table 2.2: Proposed occupational exposure limits (OELs) and derived no-effect levels (DNELs) for engineered nanoparticles.

Materials	Duration	OEL or REL (mg/m³)	DNEL (mg/m³)	Source
MWCNT (Baytubes®)	8-h TWA	0.05	–	Pauluhn (2010) [32]
MWCNT (10–20 nm/5–15 μm). Scenario NOAEC pulmonary effects	Short-term inhalation		201	Stone et al. (2010) [33]
MWCNT (10–20 nm/5–15 μm). Scenario NOAEC pulmonary effects	Chronic inhalation		33.5	Stone et al. (2010) [33]
MWCNT (10–20 nm/5–15 μm). Scenario LOAEC immune effects	Short-term inhalation		4	Stone et al. (2010) [33]
MWCNT (10–20 nm/5–15 μm). Scenario LOAEC immune effects	Chronic inhalation		0.67	Stone et al. (2010) [33]
MWCNT (Nanocyl)	8-h TWA	0.0025		Nanocyl (2009) [34]
CNT (SWCNT and MWCNT)	8-h TWA	0.007		NIOSH (2010) [35]
CNTs or CNFs	8-h TWA	0.001		NIOSH (2013) [36]
Fullerenes	Short-term inhalation		44.4	Stone et al. (2010) [33]
Fullerenes	Chronic inhalation		0.27	Stone et al. (2010) [33]
Fullerene		~ 0.8		Shinohara et al. (2009) [37]
Ag (18–19 nm)	DNEL-lung scenario 1		0.33	Stone et al. (2010) [33]
Ag (18–19 nm)	DNEL-lung scenario 2		0.098	Stone et al. (2010) [33]
Ag (18–19 nm)	DNEL-liver		0.67	Stone et al. (2010) [33]
TiO_2	0.1 risk level particles\100 nm	0.1		NIOSH (2005) [38]
TiO_2 (21 nm)	Chronic inhalation		17	Stone et al. (2010) [33]

Table 2.2 (continued)

Materials	Duration	OEL or REL (mg/m^3)	DNEL (mg/m^3)	Source
TiO$_2$ (10–100 nm; REL)	TiO$_2$ (10–100 nm; REL)	0.3		NIOSH (2011) [30]
TiO$_2$ >100 nm	TWA 10 h per day/40 days	2.4		NIOSH (2011) [30]
TiO$_2$ P25 (primary size 21 nm)	TWA 8 h a day, 1.5 days a week	1.2		Hanai et al. (2009) [39]
General	0.004 % risk level	Mass-based OEL 15		OECD (2008) [40]
General dust		3		BAuA (2009) [41]
Photocopier toner	Tolerable risk	0.6		BAuA (2008) [42]
Photocopier toner	2009 acceptable risk	0.06		BAuA (2008) [42]
Photocopier toner	2018 acceptable risk	0.006		BAuA (2008) [42]
Biopersistent granular materials (metal oxides, others)	Density >6000 kg/m^3	20,000 particles/cm^3		IFA (2009) [43]
Biopersistent granular materials	Density <6000 kg/m^3	40,000 particles/cm^3		IFA (2009) [43]
CNTs	Exposure risk ratio for asbestos	0.01 fibers/cm^3		IFA (2009) [43]
Nanoscale liquid		Mass-based OEL		IFA (2009) [43]
Fibrous	3 : 1; length 75,000 nm	0.01 fibers/cm^3		BSI (2007) [44]
CMAR		Mass-based OEL 10		BSI (2007) [44]
Insoluble	Not fibrous	Mass-based OEL 15		BSI (2007) [44]
Insoluble	Not fibrous	Mass-based OEL 10		BSI (2007) [44]
MWCNT	Bayer product only	0.05		Bayer (2010) [45]
MWCNT	Nanocyl product only	0.0025		Nanocyl (2009) [34]

CMAR carcinogenic, mutagenic, asthmagenic, and reproductive toxicants; TWA time-weighted average; CMF carbon nanofiber; CNT carbon nanotube; SWCNT single-wall CNT; MWCNT multiwall CNT; NOAEC no-observed adverse effect concentration; LOAEC lowest observed adverse effect concentration; REL recommended exposure limit. 8-h TWA 8-hour time weighted average.

Source: Adapted from Broekhuizen et al. (2012) [47].

materials by using hazard control measures and other best available practices. The following information and practices should be adopted in the work environment:

- The worker/researcher should assume that all nanoparticles are hazardous.
- To reduce their inhalation, hazardous nanoparticles should be handled in solution to prevent the generation of dust/aerosols.
- The recommended exposure limit should be minimized for the potential risk of adverse lung effects in workers who are possibly exposed at this concentration over an extended period of time.
- Health surveillance and medical screening modules should be implemented to help detect early signs of respiratory disease in workers.
- Employers must understand the risks of hazardous nanoparticle exposure and implement measures to keep all workers safe.
- Knowing how to characterize nanomaterials and processes provides clues for protection.
- To minimize and manage exposure to nanomaterials, engineering controls and personal protective equipment are the best techniques for protecting workers.
- Training programs for workers about potential hazards are crucial, and workers should have knowledge about the proper use of administrative controls, engineering controls, and safe work practices.
- Researchers must be familiar with analytical instruments and methods used to measure nanomaterial exposure levels.
- Quantitative and qualitative measurements of nanomaterial exposure are both essential for protecting workers in the workplace.

2.5.2 Exposure Monitoring

Currently, it is not clear which metrics related to the exposure to diverse nanomaterials are the most important from health, environment, and safety perspectives. Considering air contaminants, the mass-based metric has been used for characterizing toxicological effects of nanomaterials. The actual measurement of aerosolized particles that contain primary nanoparticles and agglomerates plays an important role in detecting nanomaterial emissions and evaluating control systems through field surveys. For convenience, the measurement devices used to evaluate controls in the workplace should be portable, robust, and reliable. Information about available instruments and techniques for nanoparticle monitoring has been discussed in technical reports and is summarized in Table 2.3 [36].

Table 2.3: Instruments and techniques for monitoring nanoparticle emissions in nanomanufacturing workplaces and laboratories.

Metric	Instrument	Remarks
Aerosol concentration	CPC	Real-time measurement. Typical concentration range of up to 400,000 particles/cm^3 for stand-alone models with coincidence correlation; 1,000,000 particles/cm^3 for hand-held models
	DMPS	SMPS often uses a radioactive source. FMPS uses electrometer-based sensors. Concentration range from 100–107 particles/cm^3 at 5.6 nm and 1–105 particles/cm^3 at 560 nm
Surface area	Diffusion charger	Needs an appropriate inlet preseparator for nanoparticle measurement. Total active surface area concentration up to 1000 μm^2/cm^3
	ELPI	Real-time size-selective detection of active surface area concentration. Range of 2 × 104 to 6.9 × 107 particles/cm^3 depending on size range/stage
Mass	Size-selective static sampler	Low-pressure cascade impactors. Micro-orifice impactors
	TEOM	EPA standard reference equivalent method
Aerosol concentration by calculation	ELPI	–
Surface area by calculation	DMPS	–
DMPS and ELPI used in parallel		Surface area is estimated by difference in measured aerodynamic and mobility diameters
Mass by calculation	ELPI	Calculated by assumed or known particle charge and density
	DMPS	Calculated by assumed or known particle charge and density

CPC condensation particle counter; DMPS differential mobility particle sizer; SMPS scanning mobility particle sizer; FMPS fast mobility particle sizer; ELPI electric low-pressure impactor; TEOM tapered element oscillating microbalance [36].

2.6 Conclusions

There have been enormous developments in various nanomaterials (e. g., nanotubes, nanowires, nanoparticles, nanofibers, nanowires, nanocomposites, and nanofilms), as well as in fabrication, characterization, stabilization, and commercial application worldwide. Although nanomaterials of various sizes, shapes, and structures have been produced and utilized in different consumer products, the health and environmental concerns about these nanomaterials have not been studied in detail, which creates considerable public concern. To minimize this concern, new study programs should be developed for each nanomaterial, and health and environmental concerns must be evaluated prior to industrial application of the nanomaterials. Furthermore, educational seminars, workshops, and conferences should be conducted at different universities and research centers worldwide for the awareness of students, engineers, scientists, and employees working in these fields.

Bibliography

[1] Sengul AB, Asmatulu E. Nanomaterials causing cellular toxicity and genotoxicity. In: Nanotoxicology and Nanoecotoxicology Vol. 2 (pp. 245–266). Springer, Cham, 2021.

[2] Gribbin J, Gribbin M. Richard Feynman: A life in science, Dutton, Boston, p. 170, 1997.

[3] NREL. Material science, National Renewable Energy Laboratory, Golden, 2017. https://www.nrel.gov/materials-science/scanning-electron.html (last accessed 22 Feb 2017).

[4] Sengul AB, Asmatulu E. Toxicity of metal and metal oxide nanoparticles: a review. Environmental Chemistry Letters, Sep 2020, 18(5), 1659–1683.

[5] Sahu SC, Casciano DA. Nanotoxicity: From in vivo and in vitro models to health risks, Wiley, New York, 2009.

[6] Asmatulu R. Nanotechnology safety, Elsevier, Amsterdam, 2013.

[7] Bandyopadhyaya R, Nativ-Roth E, Regev O, Yerushalmi-Rozen R. Stabilization of individual carbon nanotubes in aqueous solutions. Nano Letters, 2002, 2(1), 25–28.

[8] Wua M, Wanga Q, Lia K, Wua Y, Liu H. Optimization of stabilization conditions for electrospun polyacrylonitrile nanofibers. Polymer Degradation and Stability, 2012, 97(8), 1511–1519.

[9] Pipal AS, Taneja A, Jaiswar G. Risk assessment and toxic effects of exposure to nanoparticles associated with natural and anthropogenic sources. In: Chemistry: The key to our sustainable future, (pp. 93–103). Springer, Dordrecht, 2014.

[10] Nuraje N, Asmatulu R, Mul G. Green photo-active nanomaterials: sustainable energy and environmental remediation, RSC, Cambridge, 2015.

[11] Mehtala J, Zemlyanov DY, Max JP, Kadasala N, Zhao S, Wei A. Citrate-stabilized gold nanorods. Langmuir, 2014, 30(46), 13727–13730. https://doi.org/10.1021/la5029542.

[12] Yaszemski MJ, Trantolo DJ, Lewandrowski K-U, Hasirci V, Altobelli DE, Wise DL. Tissue engineering and novel delivery systems, CRC, Boca Raton, 2003.

[13] Kharisov B, Dias R, Kharissova OV, Vázquez A, Peñaa Y, Gómeza I. Solubilization, dispersion and stabilization of magnetic nanoparticles in water and non-aqueous solvents: recent trends. RSC Advances, 2014, 4, 45354–45381. https://doi.org/10.1039/C4RA06902A.

[14] Phosphorene. Wikipedia, last modified 14 May 2017. https://en.wikipedia.org/wiki/Phosphorene#History.

[15] Gao J, Zhang G, Zhang Y-W. The critical role of substrate in stabilizing phosphorene nanoflake: a theoretical exploration. Journal of the American Chemical Society, 2016, 138(14), 4763–4771. https://doi.org/10.1021/jacs.5b12472.

[16] NIEHS. What are nanomaterials?, 2017 National Institute of Environmental Health and Sciences, Research Triangle Park. https://www.niehs.nih.gov/health/topics/agents/sya-nano/ (last accessed 20 Mar 2017).

[17] Natário R. What is nanotechnology? networksandservers.blogspot.com, 2011. http://networksandservers.blogspot.com/2011/01/nanotechnology.html (last accessed 20 Mar 2017).

[18] Buzea C, Pacheco Blandino II, Robbie K. Nanomaterials and nanoparticles: sources and toxicity. Biointerphases, 2007, 2(4), MR17–MR172. http://arxiv.org/ftp/arxiv/papers/0801/0801.3280.pdf.

[19] Lohse S. Nanoparticles are all around us. Sustainable Nano, 2013. http://sustainable-nano.com/2013/03/25/nanoparticles-are-all-around-us/ (last accessed 12 Mar 2017).

[20] The University of Texas at Austin, Environmental Health and Safety, Nanomaterials. 2023. https://ehs.utexas.edu/working-safely/chemical-safety/nanomaterials#:~:text=EHS%20recommends%20wearing%20a%20full,glove%20box%20or%20fume%20hood.

[21] Asmatulu E, Plummer F, Miller G. Nanomaterials safety manual, Department of Environmental Health and Safety, Wichita State University, Wichita, 2014.

[22] Araki J, Mishima S. Steric stabilization of "charge-free" cellulose nanowhiskers by grafting of poly(ethylene glycol). Molecules, 2014, 20(1), 169–184. https://doi.org/10.3390/molecules20010169.

[23] Asmatulu E, Twomey J, Overcash M. Life cycle and nano-products: end-of-life assessment. Journal of Nanoparticle Research, 2012, 14, 720. https://doi.org/10.1007/s11051-012-0720-0.

[24] Falkner R, Breggin L, Jaspers N, Pendergrass J, Porter R. Consumer labeling of nanomaterials in the EU and US: convergence or divergence? Briefing paper. Chatham House, London, 2009. http://eprints.lse.ac.uk/25422/1/Consumer_labelling_of_nanomaterials_in_the_EU_and_US(LSERO).pdf.

[25] Vincent P. Nanoparticle tracking analysis (NTA): Characterization of nanomaterials for toxicological assessment. Chemistry Today, 2012, 30(6), 26–31. http://www.teknoscienze.com/Articles/Chimica-Oggi-Chemistry-Today-Nanoparticle-trackinganalysis-NTA-Characterisation-of.aspx.

[26] TSI Incorperated. Nanoparticle Monitoring in Occupational Environments – Comparing and Contrasting Measurement Metrics Nanotechnology and Occupational Health and Safety Education Series (online available 5 May 2014). https://slideplayer.com/slide/6350974/.

[27] Ziqing Z. Respiratory protection for workers handling engineered nanoparticles, 2011. http://blogs.cdc.gov/niosh-science-blog/2011/12/07/resp-nano/.

[28] Schulte PA,
Murashov V, Zumwalde R, Kuempel ED, Geraci CL. Occupational exposure limits for nanomaterials: state of the art. Journal of Nanoparticle Research, 2010, 12, 1971–1987. https://doi.org/10.1007/s11051-010-0008-1.

[29] Recommended exposure limit. Wikipedia, last modified 16 June 2013. https://en.wikipedia.org/wiki/Recommended_exposure_limit.

[30] NIOSH. Current Intelligence Bulletin 63: Occupational exposure to titanium dioxide, DHHS (NIOSH) Publication No. 2011-160, National Institute for Occupational Safety and Health, Cincinnati, 2011.

[31] Asmatulu R, Zhang B, Nuraje N. A ferrofluid guided system for the rapid separation of the nonmagnetic particles in a microfluidic device. Journal of Nanoscience and Nanotechnology, 2010, 10, 6383–6387.

[32] Pauluhn J. Multi-walled carbon nanotubes (Baytubes®): approach for the derivation of occupational exposure limit. Regulatory Toxicology and Pharmacology, 2010, 57, 78–89.

[33] Stone V, Hankin S, Aitken R et al. Engineered nanoparticles: Review of health and environmental safety (ENRHES), Edinburgh Napier University, Edinburgh, 2010.

[34] Nanocyl. Responsible care and nanomaterials case study Nanocyl, Presentation at European responsible care conference, Prague, 21–23 October 2009. http://www.cefic.be/files/downloads/04_nanocyl.pdf (last accessed 15 April 2012).

[35] NIOSH. Current Intelligence Bulletin 65: Occupational exposure to carbon nanotubes and nanofibers, DHHS (NIOSH) Publication No. 2013-145, National Institute for Occupational Safety and Health, Cincinnati, 2010.

[36] NIOSH. Current strategies for engineering controls in nanomaterial production and downstream handling processes, U.S. Department of Health and Human Services, Centers for Disease Control and Prevention, National Institute for Occupational Safety and Health, Cincinnati. DHHS (NIOSH) Publication No. 2014-102, 2013. http://ihcp.jrc.ec.europa.eu/whats-new/enhres-final-report.

[37] Shinohara N, Gamo M, Naganishi J. NEDO project: Research and development of nanoparticle characterization methods, risk assessment of manufactured nanomaterials – fullerene (C60), Interim Report 2009. New Energy and Industrial Technology Development Organization, Japan, 2009.

[38] NIOSH. Draft NIOSH Current Intelligence Bulletin: Evaluation of health hazard and recommendations for occupational exposure to titanium dioxid, November 22, 2005 National Institute for Occupational Safety and Health, Cincinnati, 2005. http://www.cdc.gov/niosh/docs/preprint/tio2/pdfs/TIO2Draft.pdf.

[39] Hanai S, Kobayashi N, Ema M, Ogura I, Gamo M, Naganishi J. NEDO project: Research and development of nanoparticle characterization methods, risk assessment of manufactured nanomaterials – titanium dioxide, Interim Report 2009. New Energy and Industrial Technology Development Organization, Japan, 2009.

[40] OECD. Working Party on Manufactured Nanomaterials: List of manufactured nanomaterials and list of endpoints for phase one of the OECD testing programme, (ENV/JM/MONO(2008)13/REV), Organization for Economic Co-operation and Development, Paris, 2009. https://one.oecd.org/document/env/jm/mono(2019)12/en/pdf.

[41] BAuA – Ausschuss fur Gefahrstoffe, Technische Regeln fur Gefahrstoffe. 900 (TRGS 900) Arbeitsplatzgrenzwerte, Bundesanstalt für Arbeitsschutz und Arbeitsmedizin, Dortmund, 2009. www.baua.de/de/Themen-von-A-Z/Gefahrstoffe/TRGS/TRGS-900.html.

[42] BAuA. Risk figures and exposure-risk relationships in activities involving carcinogenic hazardous substances, Bundesanstalt für Arbeitsschutz und Arbeitsmedizin, Dortmund, 2008. https://www.baua.de/EN/Service/Publications/Guidance/A85.pdf?__blob=publicationFile.

[43] IFA. Criteria for assessment of the effectiveness of protective measures, Institut fuer Arbeitsschutz der Deutschen Gesetzlichen Unfallversicherung, Sankt Augustin, 2009. http://www.dguv.de/ifa/en/fac/nanopartikel/beurteilungsmassstaebe/index.jsp.

[44] BSI. Guide to safe handling and disposal of manufactured nanomaterials, BSI PD6699-2:2007. British Standards Institute, London, 2007.

[45] Bayer Material Science. Occupational exposure limit (OEL) for Baytubes defined by Bayer Material Science, 2010. https://www.lawbc.com/bayer-material-sciences-announces-oel-for-baytubes/.

[46] Zhuang Z. Respiratory protection for workers handling engineered nanoparticles, 2011. http://blogs.cdc.gov/niosh-science-blog/2011/12/07/resp-nano/.

[47] Broekhuizen PV, Veelen WV, Streekstra W-H, Schulte P, Reijnders L. Exposure limits for nanoparticles: report of an international workshop on nano reference values. The Annals of Occupational Hygiene, 2012, 56(5), 515–524.

[48] Uddin M, Desai F, Asmatulu E. Review of bioaccumulation, biomagnification, and biotransformation of engineered nanomaterials. In: Nanotoxicology and Nanoecotoxicology Vol. 2 (pp. 133–164). Springer, Cham, 2021.

[49] Oberdörster G. Pulmonary effects of inhaled ultrafine particles. International Archives of Occupational and Environmental Health, 2001, 74, 1–8. https://doi.org/10.1007/s004200000185.

[50] ECHA (Europeans Chemicals Agancy). Evaluation under REACH. Progress Report. Online accessed 12/10/2009. https://echa.europa.eu/documents/10162/13628/progress_report_2009_en.pdf.

[51] Hughes S, Asmatulu E. Nanotoxicity and nanoecotoxicity: introduction, principles, and concepts. In: Nanotoxicology and Nanoecotoxicology Vol. 1 (pp. 1–19). Springer, Cham, 2021.

W. S. Khan and R. Asmatulu

3 The Importance of Safety for Manufacturing Nanomaterials

3.1 Introduction

Nanotechnology is a wide interdisciplinary field of research, development, and industrial activity that has grown rapidly worldwide over the past few years [1–3]. This field entails physics, chemistry, biology, material science, engineering and electronic processing, composites, applications, and concepts in which the defining characteristic is the size or dimension [3]. This field involves the manufacturing, processing, imaging, and application of materials that are in the size range of 1 to 100 nm. The term "nanotechnology" was first introduced in 1974 by a Japanese engineer, Norio Taniguchi. The name implies a new technology that can control materials beyond the micrometer scale [4]. To quote K. E. Drexler, "Nanotechnology is the principle of manipulation of the structure of matter at the molecular level. It entails the ability to build molecular systems with atom-by-atom precision, yielding a variety of nanomachines" [5]. Nanotechnology has the potential to change our standard of living. Some of its applications are energy storage and production, information technology, medicine, manufacturing, food, and water purification, instrumentation, and environmental uses. Several nanotechnology-based products are already available on the market, including electronic components, nanopaints, storage devices, stain-free fabrics, cutting boards, socks, and medical components. The most common nanomaterials used in consumer products are carbon nanotubes (CNTs), nanoscale metal oxides (titanium dioxide, zinc oxide), nanosilver, nanosilica, and nanogold [1–5].

The commercialization of nanotechnology has the potential to affect the health and safety of workers involved in research, the general public, customers, and the environment [1]. This emerging field has the potential for great economic expansion and is growing rapidly in various disciplines. A key aspect of this science of manipulating matter at the atomic/molecular scale is the creation of new materials at the nanoscale that have different properties compared with their bulk counterparts. For example, graphite, which is a form of carbon, is used extensively in pencils; however, when synthesized into CNTs, it becomes 100 times as strong as steel. Research in nanotechnology continues to expand around the globe and, in a few years, this field will assume a $1 trillion economy. The National Science Foundation has estimated that by 2015, nanotechnology will have a $1 trillion impact on the global economy and will employ two million workers, about one million of whom will be in the USA [1].

Acknowledgement: The authors gratefully acknowledge Wichita State University for financial and technical support of the present work.

https://doi.org/10.1515/9783110781830-003

The commercialization of nanotechnology is important; however, a complete understanding of the hazardous effects of nanoparticles/nanomaterials on human health and the environment are not available, because there has not been sufficient research in nanotechnology to answer all questions pertaining to those two areas. Furthermore, safety and health-related technologies and best industrial practices have not been imposed to protect the well-being of workers during the lifecycle of nano-based products. Safety concerns also apply to workers in nonmanufacturing enterprises, the environment, and consumers [1]. The application of nanotechnology in the medical industry for the prevention, detection, and treatment of occupational diseases (e. g., musculoskeletal disorders and pulmonary diseases) associated with exposure to nanoproducts will cost considerable sums of money. The exact cost of healthcare in this area is unknown in the long term. With the advancements in nanotechnology, new challenges, such as waste management, safety, and health risks to workers could arise in the future [2]. Progress in nanotechnology is still at mid-stage; however, it has the potential for an outstanding future in terms of improving quality of life and advancing the capabilities of materials, processes, and products in a wide range of industrial and domestic applications. Without substantial progress in establishing methodologies and relevant technologies in the workplace and laboratories, the workforce could be in great danger as a result of the unknown possible adverse health hazards of nanomaterials and nanoproducts [1].

Furthermore, airborne nanoparticles in the environment and the impact of exposure to nanoparticles must be strongly considered [1]. A strategy to deal with these issues must include the following: (i) comprehensive guidelines for health, safety, and protection throughout the life cycle of nanoproducts; (ii) research programs and focus groups to establish guidelines for the health and safety of the workforce and general public, and find ways to protect the environment from the adverse effects of nanotechnology and nanoproducts; (iii) a national research agenda related to strategy and planning for the application of nanotechnology in the biomedical industry and its long-term effects; and (iv) dissemination of the latest research findings on the life cycle of nanoproducts.

3.2 Nanotechnology Involvement

3.2.1 Scope of Nanotechnology

Nanotechnology comprises the following four areas [3]: nanofabrication, nano-medicine, nanometrology, and nanomaterials/nanoparticles. Nanofabrication is the design and manufacture of devices and systems with dimensions in nanometers. Nanomedicine is the application of nanotechnology in the medical industry. This area ranges from the medical application of nanoparticles/nanomaterials in nanobiosensors as well as the possible application of molecular nanotechnology and nanobiotechnology. Nanometrology is concerned with the science of measurement at the nanoscale. Particles having an aerodynamic diameter of less than 100 nm are regarded as true nanoparticles [6]. Nanoparticles with novel physicochemical properties are used to improve the functionality of commercial and consumer goods [6]. Examples of such products are paints, sunglasses, sunscreen, cosmetics, building materials, clothing, electronics, solar cells, industrial lubricants, semiconductors, advanced tires, and fuel cells [6]. Titanium dioxides are used in paints, and zinc oxides are used in sunscreen products. Nanoparticles provide great benefits in almost all sectors. However, the properties of particles that are scientifically and commercially exploitable could be the basis of some adverse health and environmental effects. There is a growing need to identify those nanoparticles with excellent commercial potential, evaluate techniques for their manufacture, conduct hazard assessments of those materials, and analyze their exposure assessments. Currently, limited information is available on human exposure and the environmental effects of nanomaterials. Nanotechnology will continue to grow, with more and more applications emerging globally in the near future. Thousands of workers may be potentially exposed to engineered nanoparticles as a result of the recent acceleration in their manufacture and application. Research on the toxicity of engineered nanoparticles has demonstrated greater biological activity of nanoparticles compared with large particles of the same material composition. Laboratory tests have reported significant toxicity in animals exposed to nanomaterials. Determining the kind of adverse health effects that nanomaterials pose to human health in the workplace is a crucial issue.

Nanomanufacturing is the bridge between the innovation and discovery of nanoscience and real-world nanotechnology products. Through nanomanufacturing, the potential of technological innovations across a spectrum of products will ultimately affect all industrial sectors in the near future. Nanomanufacturing is the fabrication of nanoscale building blocks (nanomaterials, nanostructures) into higher-order structures, such as nanosystems and nanodevices, and integrating these into larger structures and systems. Nanotechnology is generally considered a technology that uses the "bottom-up" approach for creating materials, devices, and systems, in contrast to the traditional "top-down" approach [4]. The manufacture of nanocomponents and nano-products involves a wide variety of synthesis technologies [7]. These techniques can be classified on the basis of approach (top-down or bottom-up) and the nature of the synthesis (wet or dry).

A top-down approach for nanomanufacturing involves the creation of nanoproducts and nanocomponents using bulk material as a starting block. Examples of such techniques are lithography and etching [7]. Lithography is a process that allows the patterning of a required design onto the starting material. This technique is used to produce the miniaturization of electronic components, such as computer chips, CDs, and DVDs. Etching is a process used to create a precision surface and the attrition of metals to create metal nanoparticles [7].

In the bottom-up approach, nanocomponents are built up from the atomic/ molecular level. An example of this kind of approach is found in nature; cells use enzymes to produce DNA by binding molecules together to make the final structure.

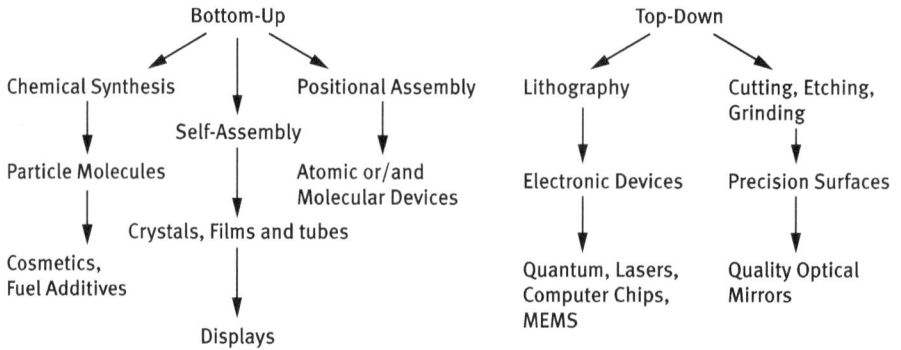

Figure 3.1: Bottom-up and top-down techniques used in nanomanufacturing [1].

Other examples of this approach are self-assembly and chemical synthesis. Cosmetics, additives in fuel, and atomic/molecular devices are examples of products manufactured using the bottom-up approach. Figure 3.1 illustrates some of the types of materials and products used in the bottom-up and top-down approaches [1].

The primary requirements in nanomanufacturing are that the manufacturing involves a dimensional scale of 1–100 nm, and that the processes and resulting nanostructures exploit the physical phenomena unique to that scale. Nanomanufacturing begins with nanoscale building blocks, such as quantum dots, buckyballs, nanotubes, and nanowires, and their production in volume. Nanoscale building blocks are then integrated into nanostructures. In turn, nanostructures are used to manufacture nanodevices, which are used to form nano subsystems, which are then integrated into nanosystems. From here, nanosystems form useful products that are inserted into large-scale systems and platforms, such as cell phones, power grids, and airplanes.

3.2.2 Nanotechnology Education and Research Programs

Educational and research programs in nanotechnology are greatly needed in order to determine the impacts of nanotechnology products, as well as occupational and environmental health and safety (OEHS) issues. The following two perspectives should be considered, as shown in Figure 3.2 [1]:

- Protection of an individual's safety and health throughout the life cycle of nano-products.
- Application of new technology for the prevention and detection/treatment of occupational and environmental maladies/diseases.

Figure 3.3 shows the nanoproduct life cycle, which mainly consists of nanomanufacturing (processing, materials handling, manufacturing, environment, and supply chain), nanoproducts (customer and industry use), and after-use (disposal and discharge).

Nanotechnology
Interdisciplinary Field

Promotion and Prevention of
Public/Workers/Consumers
On Safety and Health Hazards
of Nanoproducts

Use of Nanotechnology in
Prevention and
Detection/Treatment of
Occupational and
Environmental Diseases

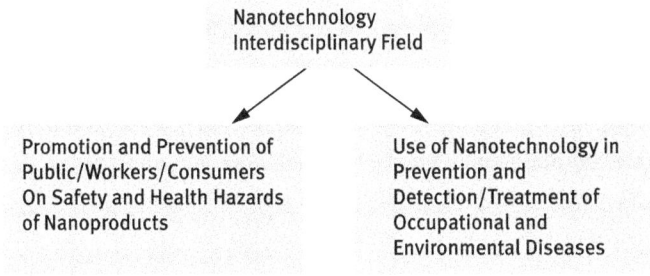

Figure 3.2: Objectives of nanotechnology from an occupational and environmental health and safety point of view [1].

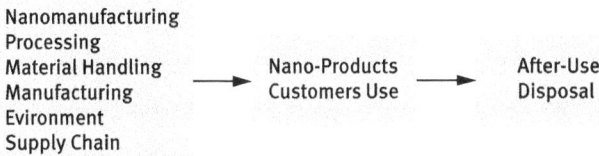

Nanomanufacturing
Processing
Material Handling   Nano-Products   After-Use
Manufacturing   Customers Use   Disposal
Evironment
Supply Chain

Figure 3.3: Life cycle of nanoproducts [1].

Universities, research centers, government institutions, and private companies should be involved in every step of nanoproduct production. An education program should address issues such as training the workforce in handling nano-based products, as well as finding solutions to minimize waste and exposure to nanomaterials, nanoparticles, and nanoproducts, thus creating an atmosphere of awareness of the hazardous effects of nanomaterials in the workforce.

Education Program: An educational program should consist of four major steps [1]:

– Determine the potential health hazards associated with exposure to nanomaterials/nanoproducts by reviewing the literature available regarding this issue.
– Derive qualitative and quantitative exposure estimates associated with nanomanufacturing by reviewing the literature regarding exposure to nanomaterials, international exposure standards, and current technologies to minimize airborne nanoparticles; develop promotion and protection intervention for a workforce well-being and apply this learning.
– Address the issue of safety and potential health hazards in nanomanufacturing enterprises in all manufacturing sectors and educate the workforce on safety and health hazard issues associated with exposure to nanomaterials.
– Identify issues related to safety and health hazards associated with the exposure to the general public and consumers of nanoproducts to nanomaterials. Although nanotechnology is expected to increase life expectancy and enhance product capability, many uncertainties are still associated with the exposure to the environment

of nanoproducts, and related health issues. An educational program designed to address these issues would be valuable.

Research Program: A research program aimed at protecting the workforce from the hazardous effects of nanomaterials and nanodevices would have two main objectives [1]:

– Provide a platform for the exchange of ideas and research outcomes on the hazardous effects of nanomaterials on the environment and human health. The aim would be to develop a national agenda for assessing, the impact of nanotechnology on occupational health, along with the life cycle of nanoproducts and their disposal techniques.
– Provide a platform for the exchange of ideas and research outcomes on the use of nanotechnology for the prevention, early detection, and treatment of specific diseases associated with exposure to nanoproducts.

3.3 Nanostructured Materials

Nanostructured materials are products of nanotechnology and have enormous potential in various fields of materials science, engineering, technology, and biomedical science [8]. Nanostructured materials can exhibit outstanding physical and chemical properties due to their quantum size, high surface area, high aspect ratio, few defects, and macro quantum tunnel effects [8]. The rapid growth of nanotechnology suggests that it will not take long before a wide variety of new electronic, pharmaceutical, and other industrial uses for nanostructured materials are found [9]. Such rapid proliferation of nanotechnology has prompted great concern over the safety and environmental effects of nanostructured materials [9].

Nanotechnology has revolutionized our society. Nanomaterials are being used in almost all industries, and their usage is increasing in an ever-growing number of products and applications. Such rapid advancement of nanomaterials has stimulated the demand for a safety assessment with respect to both human health and the environment. The potential advantages of these materials are too numerous to count, and industries have great expectations, but there are many unanswered questions regarding the safety of workers and potential hazards to humans and the environment following exposure to nanomaterials. It is very difficult to classify these nonhomogeneous materials, because they exist in various forms, and their impact on human health and the environment is still unknown. There are some daunting analytical problems associated with characterizing nanomaterials into different categories, assessing exposure to these materials, and elucidating their potential pathways. The lack of such information related to nanomaterial exposure is impeding the potential risk assessment and life cycle assessment (LCA) of nanomaterials. Some LCA methods are being used by researchers to assess the potential hazardous impact of nanomaterials, but the assessment is still in its early stages and information is limited.

3.3.1 Nanoparticles

Nanoparticles are the end products of a variety of chemical, physical, and biological processes [10]. A nanomaterial is a substance with at least one dimension less than 100 nm and can take many different forms, such as particles, wires, tubes, rods, and spheres. Nanoparticle products include quantum dots, iron oxides, titanium dioxides, aluminum oxides, cerium oxides, zinc oxides, silicon dioxides, gold nanoparticles, dendrimers, and some layered structures. Recently, there has been rapid growth in the development of new nanoparticles. One edition of the *Journal of Material Chemistry* included 47 papers on the development of nanoparticles [9]. Table 3.1 shows the main categories of nanoparticles according to their morphologies, applications, and composition [9].

Table 3.1: Different categories of nanoparticles available on the market [9].

Nanostructures	Materials
Fullerenes	Carbon
Nanotubes	Carbon, boron nitride
Nanowires	Metals, semiconductors, oxides, sulfides, nitrides
Nanocrystals	Insulators, semiconductors, metals
Quantum dots	Magnetic materials
Other nanoparticles	Iron, zinc, titanium and ceramic oxides, metals/alloys, composites, polymers, etc.

3.3.1.1 Fullerenes

Fullerenes are a naturally occurring form of carbon, discovered in 1985 [9]. They are a series of carbon molecules that form either a closed hollow sphere (buckyballs) or a cylinder (nanotube). Fullerenes are similar in structure to graphite, which is composed of a sheet of hexagonal rings that form a three-dimensional structure [9]. Fullerenes are produced by laser ablation of graphite in a helium atmosphere [9]. The smallest structure is a cage-like molecule composed of 60 carbon atoms (C_{60}) joined together by single and double bonds to form a hollow sphere with 12 pentagonal and 20 hexagonal faces. Other techniques used to produce fullerenes are the combustion of hydrocarbon, thermal and nonthermal plasma pyrolysis of coal and hydrocarbons, and thermal decomposition of hydrocarbon [9].

3.3.1.2 Carbon Nanotubes

Carbon nanotubes (CNTs) were first discovered by Iijima in 1991 in Japan using an arc-discharge method. CNTs are a special form of fullerene, consisting of concentric layers of graphite (multiwalled CNTs, MWCNTs). However, nanotubes composed of a single layer (single-walled CNTs, SWCNTs) were discovered during the analysis of ash from a synthesis reactor. CNTs are similar in structure to C_{60} (buckyballs), but they are elongated to form a tubular structure [9]. A SWCNT has a diameter of 0.6–5 nm, whereas a MWCNT has an inner diameter of 1.5–15 nm and an outer diameter of 2.5–50 nm. CNTs can be produced in various aspect ratios and varying lengths, depending on the processing technique [9]. CNTs possess outstanding mechanical, thermal, and electrical properties [11]. Their mechanical strength is 150 GPa, thermal conductivity is 1500–300 W/m K, and electrical conductivity is as high as 104 S/cm [11]. CNTs are 100 times stronger than steel, and their thermal conductivity is 4–5 times higher than that of most metals. They are potentially one of the strongest materials known to date [9].

3.3.1.3 Nanowires

Nanowires are nanostructures composed of either conducting or semiconducting nanoparticles with diameters of 1–100 nm and large aspect ratios. Nanowires are used as interconnectors in nanodevices [9]. At the nanoscale, quantum effects become predominant; therefore, these wires are also known as "quantum wires." Various types of conducting and semiconducting nanowires have been fabricated, including nickel, gold, platinum, silicon, and, recently, gallium nitride (GaN). However, some insulating nanowires, including zinc oxide (ZnO), stannic oxide (SnO_2), silica (silicon dioxide, SiO_2), and titanium dioxide (TiO_2), have also been produced.

3.3.1.4 Quantum Dots

Quantum dots are very tiny particles or nanocrystals of semiconductor materials, metal oxides, or assemblies of metals. They have diameters of 2–10 nm and exhibit novel optical, electronic, magnetic, and catalytic properties [9]. The unusual properties of quantum dots are the result of the high surface-to-volume ratios of these particles. Quantum dots are neither an extended solid structure nor a single molecular entity [9]. Various techniques are available to produce them; however, the wet chemical colloidal process is the most common technique [9]. Scientists have applied quantum dots in solar cells, light-emitting diodes (LEDs), transistors, and medical imaging.

3.3.1.5 Metallic Nanoparticles

A nanoparticle is the basic component of a nanostructured material. Generally, the size of a nanoparticle is in the range of 1–100 nm. The term "metallic nanoparticle" is used to describe nanosized metal with dimensions (length, thickness, or width) in this range. Metallic nanoparticles exhibit different physical and chemical properties from their bulk counterparts, and some of these properties might prove attractive in industrial applications [28]. Nanoparticles possess some unique features, such as high surface-area-to-volume ratio, large surface energy, quantum confinement, and short-range ordering. Commercially available metallic nanoparticles include Ag, Au, Pt, ZnO, and metal oxides, such as CuO, SiO_2, TiO_2, alumina (Al_2O_3), and iron oxides (Fe_3O_4, and Fe_2O_3).

3.3.1.6 Carbon Black

Carbon black is produced by the incomplete combustion of fossil fuels. Generally, anthropogenic combustion produces a wide variety of particles, including some ultrafine particles, which are compatible with the definition of nanoparticles [12]. These particles are referred to as carbon black and are the result of the incomplete combustion of heavy petroleum products. Carbon black is used as a reinforcing material in automobile tires and rubber products, paints, and color pigments. The particle size for carbon black is in the range of 10–300 nm [12].

3.3.1.7 Dendrimers

Dendrimers are highly branched, monodispersed macromolecules with a star-like appearance and nanosized dimensions. Dendrimers have three components: a central core, an interior structure (branches), and an exterior surface with functional groups. By varying these three components, dendrimers of different shapes and sizes can be produced. Their structure greatly impacts their physical and chemical properties. Dendrimers are an ideal candidate for applications in biology, engineering, and material science.

3.3.1.8 Nanocomposites

Nanocomposites are materials to which nanosized filler components are added to improve the properties of the resulting materials. Nanocomposites are composed of two or more distinct constituents or phases having different physical and chemical properties and are separated by a distinct interface. Their many unique properties are not depicted by any of the constituents. The constituent that is generally present in greater quantity is called the matrix. The constituent that is embedded into the matrix material to improve the mechanical properties of nanocomposites is called the reinforcement (or nanomaterial). Reinforcement is generally in the form of nanosized filler material. Generally, nanocomposites show anisotropy (properties are directionally dependent) because of the distinct properties of the constituents and inhomogeneous distribution of the reinforcement.

3.3.1.9 Nanoclays

Nanoclays, a modified form of layered mineral silicates, are a class of organic-inorganic hybrid material [13]. Depending upon the chemical composition and morphology of the particles, nanoclays can be categorized into many classes, such as montmorillonite, bentonite, kaolinite, hectorite, and halloysite. These materials are finding a wide range of applications in polymer nanocomposites, absorbents for toxic gas emissions, drug delivery carriers, rheological modified inks, fire-retardant materials, paints, and greases [13]. "Layered silicate" is a generic term referring to synthesized layered silicates (montmorillonite, laponite, and hectorite) and natural clays [13]. Montmorillonite is the most common nanoclay used in many materials applications. The plate-like montmorillonite consists of a 1-nm thick aluminum silicate surface layer modified with cations having dimensions (length and width) of hundreds of nanometers [13]. Nanoclays are naturally occurring materials that contain nanosized particles within the mined material.

3.3.1.10 Nanocrystals

Nanocrystals are crystalline particles having nanometer dimensions. Nanocrystals are the building blocks of nanotechnology. Their properties can be changed by controlling the methods of synthesis. They can be incorporated into electronic devices, such as LEDs for energy-efficient lighting and are being used in filtration to refine crude oil into diesel fuel. Nanocrystals are also finding applications in many other areas, such as solar cells, catalysts, and sensors.

3.4 Toxicity of Nanomaterials

Nanomaterials can come in contact with the vascular endothelium and cause cardiovascular damage [14]. Generally, nanomaterials enter the human body via inhalation, dermal, or oral routes. Some researchers have found that nanomaterials can promote DNA damage. Previous studies have shown that nanoparticles can cause some level of damage to cells. Silver nanoparticles are more toxic, and endothelial cells are more susceptible than bulk silver materials when exposed to silver nanoparticles. Titanium dioxide nanoparticles are less toxic than silver particles. Nanomaterials possess unique toxic properties due to their physicochemical characteristics and nanosize features [14]. Some nontoxic microparticles become toxic when they are reduced to the nanoscale. Nanomaterials are more toxic than bulk materials of the same chemical composition.

Nanoparticles are very reactive and possess catalytic properties as a result of their high surface-area-to-volume ratio. They have access to transport mechanisms that are not possible for large structures, thus bypassing all biological barriers and penetrating into the interior of cells [14]. Therefore, significant attention has been paid to nanoparticle toxicity, because the impact of many nanomaterials on human health is not yet known. There is considerable concern regarding the toxicity of nanomaterials and the methods of toxicity assessment. Toxicity evaluation of nanomaterials provides some insights into the adverse effects of nanomaterials and helps in developing a database that can provide useful information for the assessment of potential risk and risk-management issues [14].

The latest studies show that the potential genotoxic, oxidative, and cytotoxic effects at the cellular level, as well as respiratory, neurotoxic, cardiovascular, dermal, and immunological effects, can be caused by exposure to different nanomaterials [14–16]. To determine the toxicity of nanomaterials, both in vitro and animal studies are being used. Generally, an in vitro assay consists of subcellular systems, cellular systems, individual cells, and tissues. Although in vitro data is not a substitute for whole-animal studies, it helps in establishing a platform for further assessment of the potential risk of the hazardous effects of nanomaterials [16]. The results of in vitro studies can be extrapolated to human health. Toxicity data from in vitro studies can be used to screen acute hazards and possible mechanisms of compound interactions with animals or humans. Advancement in toxicity testing aims at the characterization, uptake, and mechanisms of toxicity of nanomaterials in all types of cells. After characterization, the interactions of nanomaterials with cells can be determined by biochemical assay and microscopy. Viability testing, combined with studies of nanomaterial morphology and the generation of oxidative stress, is helpful in elucidating the toxicity mechanism. These tests provide useful information for determining how the size of nanomaterials, their chemical composition, and their functionalization contribute to toxicity [16].

3.4.1 Toxicity of Carbon-Based Nanomaterials

The most widely used nanomaterials are based on carbon and include fullerenes, SWCNTs, MWCNTs, carbon black, and graphene. Some studies have demonstrated the pulmonary toxicity of SWCNTs in rats [17]. It has also been determined that SWCNTs cause dose-dependent interstitial granulomas and pulmonary injuries and are more toxic than

quartz. Basic approaches generally used to determine the toxicity of carbon nanomaterials include the following [17]:
- Assay for inhibition of mitochondrial dehydrogenase activity
- Phagocytic response to latex beads
- Transmission electron microscopy (TEM)

3.4.1.1 Assay for Inhibition of Mitochondrial Dehydrogenase Activity

Assay for inhibition of mitochondrial dehydrogenase activity is a common method in cytotoxicity testing. Alveolar macrophage (AM) cells are the first line of immunological defense against nanoparticles in the lungs [17]. For this test, adult pathogen-free healthy pigs are chosen and AM cells obtained using the bronchoalveolar lavage medical procedure [17]. AM cells are isolated from the lung lavage fluid and then suspended in RPMI (Roswell Park Memorial Institute) 1640 medium containing 10 % FBS (fetal bovine serum). The macrophages are plated at a specified density of viable cells per well, in a coaster in order to attach the plastic matrix. Afterward, the medium is removed, and a fresh cell monolayer is exposed to the nanomaterial [17]. The cytotoxicity of nanomaterials is determined by the MTT colorimetric assay. This assay is used for quantification of cell death and cell lysis and is based on the measurement of lactate dehydrogenase (LDH) activity released from damaged cells into the supernatant. After exposing AM cells to the nanomaterial, phosphate-buffered saline (PBS) is added to each sample. Incubation for a specified period of time results in dark formazan crystals. These crystals are dissolved in HCl–2-propanol, and then centrifuged to remove any traces of particles. The supernatant is realiquoted into a new well plate, and the absorbance is recorded using a Biorad Microplate Reader [17]. Four steps must be performed for calculating the percentage cytotoxicity, as follows [18]:
- *Background control*: Provides information about the LDH activity contained in the medium.
- *Low control*: Provides information about the LDH activity released from untreated cells.
- *High control*: Provides information about the maximum release of LDH activity from cells.
- *Experimental value*: Provides information about the LDH activity released from treated cells.

To determine the percentage cytotoxicity, the average absorbance values of triplicate samples are subtracted from each absorbance value obtained for the background control. The resulting values are then substituted into the following equation [18]:

$$\text{Percentage cytotoxicity} = \frac{\text{Experimental value} - \text{low control}}{\text{High control} - \text{low control}} \times 100 \qquad (3.1)$$

3.4.1.2 Phagocytic Response to Latex Beads

This technique is used to demonstrate the immunological function of the alveolar macrophage [17]. The phagocytic ability of isolated primary AM cells exposed to carbon nanomaterials can be assessed by measuring their ability to phagocytose colloidal gold latex breads. After exposing AM cells to various doses of nanomaterials, they are transferred to fresh medium containing latex breads. Beads that are not phagocytosed are removed from the medium and washed with PBS solution. Then, cells are tested under a fluorescent microscope. The adhered particles and phagocytized particles can be counted separately by adjusting the focal length in the microscope [17].

After microscopic observation, cells are harvested and investigated using a flow cytometer. From the plot of forward scatter versus side scatter, the AMs can be extracted and cells isolated. Then, free particles can be distinguished on the basis of granularity and size [17]. The phagocytic ability can be expressed as the geometric mean of the fluorescent intensity of the phagocytized beads compared with that of total beads [17].

3.4.1.3 Observation Using Electron Microscopy

Electron microscopy is a well-known technique for nanoimaging particles of a small size to show their morphology, and material features [19]. Many studies depend upon TEM for information on particle size, shape, morphology, and aggregation. TEM is generally used for imaging metallic samples; however, energy-filtered TEM (EFTEM) can be used to image nonmetallic samples [19]. Structural alteration of AM cells induced by nanomaterials is generally observed by TEM. AM cells are harvested using a cell scraper, washed with PBS solutions, and prefixed with glutaraldehyde [17]. After washing, the AMs are postfixed with osmium tetroxide and again washed with a cacodylate buffer. The cells are then analyzed with TEM after dehydration, ultrathin sectioning, and staining with uranyl acetate and lead citrate [17].

3.4.2 Toxicity of Metal-Based Nanomaterials

Nanoparticles can cause harmful effects on tissues and organs, as well as at the cellular, subcellular, and protein levels in the body, as a result of their abnormal physio-chemical properties [23]. Metal nanoparticles have attracted significant attention because of their wide range of applications in the medical, consumer, industrial, and military sectors. As the particle size decreases, some metal nanoparticles exhibit toxicity, even though the same material is inert in its bulk form (e. g., Au, Ag, and Pt). Metal nanoparticles interact with enzymes and proteins, and also interfere with the antioxidant defense mechanism, resulting in the generation of reactive oxygen species, initiation of the inflammatory response, and perturbation and destruction of mitochondria, causing apoptosis [23].

To investigate the health consequences of exposure to metal nanoparticles, an in vitro system is generally used to predict effects at the cellular level. Toxicity testing of nanomaterials is used for characterization, uptake, and the mechanism of toxicity in a variety of cell types. After the successful characterization of nanomaterials, the interaction of nanomaterials with cells can be studied with the help of biochemical assays and microscopy techniques. Results of microscopy observation, viability testing, and the generation of oxidative stress can help in elucidating the mechanism of toxicity.

Silver is not normally found in sufficiently high concentrations to pose a real threat to humans and the environment; however, nanosilver particles have surface and physical properties that could pose a threat to both [24]. The release of toxic silver ions from nanosilver particles is of great concern. Such particles exhibit high toxicity as a result of the activity of free silver ions released from nanoparticles. Silver nanoparticles can cause chromosomal aberrations and DNA damage. Some studies have shown that silver nanoparticles can enter into cells and cause cellular damage [24]. Many in vitro studies have been conducted on the adverse effects of silver nanoparticles with a size of 1–100 nm. The uptake of silver nanoparticles by different cells has been reported in many in vitro studies. Most related publications show reduced cell viability after exposure to silver nanoparticles. Some in vitro studies show glutathione depletion, mitochondrial derivations, and damage to cell membranes. Exposure of human peripheral blood mononuclear cells to nanosilver causes inhibition of phytohemagglutinin (PHA)-induced proliferation [24]. The toxicity of silver nanoparticles is higher than that of most carbon-based nanomaterials and many metal nanoparticles [16]. The toxicity of silver nanoparticles increases with a decrease in size and increase in concentration, as a result of oxidative stress. Research studies have shown that C18-4 germline stem cells are more sensitive to 15 nm silver nanomaterials than are BRL 3A liver cells and CRL-2192 alveolar macrophages, after 24 hours of exposure to 15 nm silver nanoparticles [16].

A large number of other metal nanoparticles have been screened to determine their toxicities using assays that reveal LDA leakage through the plasma membrane [16]. The LDA release of BRL 3A rat liver cells was determined after 24 h exposure to different nanoparticles, showing that exposure to microparticles of cadmium oxide increased membrane leakage [18]. Silver nanoparticles exhibited higher LDH leakage at concentrations of 150 µg/mL than metal oxide nanoparticles. The evaluation of nanomaterial uptake into cells is useful in determining the toxicity of nanomaterials [16] and can be monitored using fluorescent microscopy, flow cytometry, and fluorescence-activated cell sorting. Some advanced techniques, such as ultrahigh-resolution light microscopy and wet imaging under a high vacuum are also being used, in addition to TEM observation on thin films [16].

Titanium dioxide nanoparticles are commonly used because of their photocatalytic properties, availability, and low cost [25]. Some studies show the biokinetic activity of TiO_2 nanoparticles both in vitro and in vivo. However, it is difficult to compare results, because TiO_2 exists in different crystalline phases, sizes, and shapes. Some studies indicate that it can be readily uptaken by A549 cells (carcinomic human alveolar basal epithelial cells) in vitro. However, absorption through the differentiated Caco-2 monolayer system (human epithelial colorectal adenocarcinoma cells) and in an in vivo oral study was low [25].

3.5 In Vitro Assessments of Nanomaterial Toxicity

The significant expansion of technological and commercial interest in nanomaterials has stimulated the field of nanotechnology. Around 1,500 consumer products of nanotechnology are available on the market [19, 26]. The global market requires tons of raw nanomaterials, ranging from nanoscale metals and metal oxide particles to carbon-based nanomaterials [19]. This robust manufacturing and consumer usage produce many sources that release nanomaterials into the environment, water, food supplies, ecosystem, and other routes of entry into the human body [19]. When a material's dimensions approach the nanoscale, certain properties, such as capillary force, melting point, optical characteristics, conductivity, surface energy, magnetism, electron affinity, polarization ability, reactivity, and ionization potential can become scale-dependent [19, 27].

This section outlines the methods generally used to assess the surface and bulk properties and biological reactivity of nanomaterials in a model system in an in vitro system. These assays are significant in order to characterize nanomaterial applications in biotechnology, ecosystems, biomedicine, and cytotoxicity screening. There is no consensus about the risk, toxicity, hazards, and environmental effects of almost all nanomaterials. At present, the tools available to determine the pharmacological and toxicological characteristics of nanomaterials are too primitive to provide much information [19].

3.5.1 Detection of Surface Contamination

Surface contaminants can be detected by many widely used techniques, such as time-of-flight secondary ion mass spectroscopy (ToF-SIMS), X-ray photoelectron spectroscopy (XPS), X-ray fluorescence (XRF), energy-dispersive X-ray analysis (EDX), and surface-enhanced Raman spectroscopy (SERS) [19].

ToF-SIMS is a surface analytical technique that focuses a pulsed beam of primary ions onto a sample surface, producing secondary ions in a sputtering process. Analyzing these secondary ions provides information about the molecular and elemental species present on the surface of the sample [19]. XPS is a widely used surface analysis technique, whereby a sample is irradiated with a beam of X-rays under an ultra-high vacuum. The kinetic energy and number of electrons that escape from the top surface of the sample are analyzed [19]. XRF is a technique in which X-rays or gamma rays are bombarded onto a sample. The emission of secondary fluorescent X-rays photons yields quantification of all elements present in the sample [19]. EDX irradiates the sample with X-rays and measures the energy (wavelength) and intensity of the X-rays emitted from the sample to determine the composition of the sample. This technique scans only 5 µm on the top of the sample; however, it can provide an entire spectrum simultaneously. SERS is a surface-sensitive technique, in which an intense laser beam is focused on the surface. Localized surface plasmons are excited due to the vibration on the sample surface and leave the sample with a different frequency [19]. The energy change reflects the chemical composition of the sample surface.

3.5.2 Particle Sizing and Aggregation

Nanomaterials possess a very high surface and low diameter, so all physical, chemical, physicochemical, and biological properties blend together because of their small size. Thus, nanoparticle sizing is a crucial aspect of pre-characterization [19]. Moreover, because of the high dispersion and high surface energy of nanoparticles, aggregation is an issue when handling nanoparticles experimentally. At the nanoscale, aggregation is extremely difficult to avoid. Several properties, such as colloidal stability, homogeneity, optical and electronic features, and cell uptake are adversely affected by aggregation. Widely used techniques for characterization include scanning electron microscopy (SEM), TEM, dynamic light scattering (DLS), optical spectroscopy, and fluorescence polarization [19].

3.5.2.1 Transmission Electron Microscopy

TEM is a well-known technique for micro- and nanosized imaging of material features, particle size distribution, surface morphology, and aggregation [19]. Imaging of particles by TEM provides sizing information through direct electron imaging, but it is usually used for metallic samples. However, EFTEM can be used for imaging non-metallic samples. TEM uses electrons as a "light source" for transmission through the specimen. Electrons interact with the specimen as they pass through and form an image, both inside and out. The image is magnified and focused onto a fluorescent screen or a detector. The information obtained by TEM for solution aggregation may be inconclusive and should be confirmed by other techniques such as zeta-potential spectrophotometry and gel electrophoresis [19]. Many samples, upon drying, produce aggregation as a result of increased ionic strength and surface tension. TEM analysis cannot control the effects of surface tension. Information obtained by TEM in ex-situ conditions may not necessarily be a true representative of in situ aggregation states, because preparation of a nanoparticle solution for TEM imaging requires sample desiccation [19].

3.5.2.2 Scanning Electron Microscopy

SEM uses electrons to impinge on the surface of a specimen. The electrons interact with the atoms on the surface of the specimen, producing signals that can be detected and contain information about the specimen's surface features and composition. SEM requires a high vacuum and a dried specimen, which causes some experimental uncertainty and the inability to produce in situ characteristics accurately [19]. However, modern SEM can scan and image the specimen in hydrated conditions under a high vacuum.

3.5.2.3 Optical Spectroscopy

Optical spectroscopy is the measurement of the interaction of light with the specimen. Some metallic nanoparticles display size-dependent absorption and scattering of incident light through excitation of the metal's plasmon band electrons by incident photons or through a scattering of incident photons [19]. Free electrons and electronic coupling of metal lattice energies, and excitation interband energies are essential for plasmon excitation. These requirements are found in a few metals, such as lead, mercury, tin, cadmium, and gold. In most metals, the plasma frequency is in the ultraviolet region of the electromagnetic spectrum. Some metals, such as gold and copper, have interband transitions in the visible range of the electromagnetism spectrum, resulting in the absorption of specific light energies (color), thereby yielding the color of interest [19].

Other metals and metal oxides display plasmon in a nonvisible region, thereby making plasmon excitation and detection extremely difficult. Metal oxide formation and intrinsic surface metal lattice mismatch generally distort the energy interband coupling. Furthermore, the free-electron phenomenon diminishes the plasmon coupling effects in most metals, except those that are oxide-free (noble metals) [19]. Shifts in surface plasmon band excitation take place in metal as a result of adsorbate binding, which causes changes in the surface interband electronic states. Thick surface-stabilizing layers and surface oxide can prevent other adsorbate electronics. Gold and silver nanoparticles are generally sized by measuring the extinction wavelength of incident light [19]. Plasmon absorbance decreases in intensity and red-shifts to higher optical wavelengths as the average particle diameter increases. This results in characteristic plasmonic peaks for each size of metal nanoparticles. Likewise, adsorption of contaminants, stabilizing layers, and DNA on the nanoparticle surface red-shift the extinction wavelength by a few nanometers. This shift indicates particle aggregation. Gold colloidal solutions change in color from red to blue, thus indicating particle aggregation [19].

Optical spectroscopy has been used in a wide variety of analytical techniques. The method can selectively detect and identify a large number of substances and serve as an important tool in the pharmaceutical and chemical industries. Recently, optical spectroscopy has been used to detect and identify engineered nanomaterials.

3.5.2.4 Dynamic Light Scattering

Particle size can also be determined by measuring changes in the intensity of light scattered from a solution. This technique is generally called dynamic light scattering (DLS) but is also known as photon correlation spectroscopy (PCS). Applications of this technique are the characterization of particles, molecules, and emulsions in solution.

In this technique, a beam of monochromatic light strikes a solution containing spherical particles. Brownian motion of the particles causes a Doppler shift, changing the wavelength of the incoming light [20]. Brownian motion of the particles in suspension causes light to be scattered at different intensities; thus, by analyzing the fluctuation in intensity, it is possible to measure particle size distribution, particle motion in solution, and the diffusion coefficient of the particles [20]. This technique is cost-effective, automatized, and easy to operate.

3.5.2.5 Fluorescence Polarization

Fluorescence polarization (FP) is based on the fact that a fluorescent molecule in solution is excited by a plane-polarized light and emits polarized fluorescent light back into a fixed plane if the molecules remain stationary during excitation and emission [21]. Molecules rotate and tumble, and the planes into which light is emitted can be different from the excitation plane. The polarization of a molecule is related to the molecule's rotational relaxation time (molecule rotation through an angle of 68.5°). The rotational relaxation time depends on absolute temperature, molecule volume, gas constant, and viscosity [21]. Recently, time-resolved fluorescence polarization anisotropy (TRFPA) has been used in nanosystems for measuring particle size. The fluorescence polarization decay time is related to particle size in accordance with the Stokes–Einstein–Debye rotational equation for particle motion. With this method, particles having a size of 1–10 nm can be measured with 0.1 nm resolution [19].

3.5.2.6 Other Techniques

Several other techniques are available for quantifying the sizing of nanomaterials, as follows [19]:
- X-ray diffraction (XRD) is used to differentiate between crystalline and nanocrystalline samples.
- Multiangle laser light scattering is an analytical technique for finding the absolute molar mass, size, and structure of macromolecules and particles in solutions. This technique is used in combination with ultraviolet-visible spectroscopy and field-flow fractionation for particle sizing.

- Small-angle X-ray scattering, and small-angle neutron scattering are used for studying particle size and shape.
- Inductively coupled plasma–mass spectroscopy (ICP-MS) is a type of mass spectroscopy that is used to measure/detect metals and several nonmetals.
- Inductively coupled plasma–atomic emission spectroscopy (ICP-AES) is used for the detection of trace metals.

3.6 Nano-safety

The National Institute for Occupational Safety and Health (NIOSH) is a US federal agency mainly responsible for supervising research and preventing work-related injuries and illnesses. NIOSH is also responsible for protecting workers from injuries and illness in the workplace in the future, which is important, because the applications of nanotechnology are advancing rapidly [2, 22]. Nanotechnology has a wide range of applications, with some unknown safety and health risks. Because nanotechnology is a new field and the potential risk and hazardous effects of nanomaterials are unknown, precautionary measures must be taken [2, 22]. There are many concerns and uncertainties related to the use of engineered nanomaterials and whether their exotic properties pose any adverse threat to human health and the environment. The data currently available to determine the adverse effects of engineered nanomaterials is not adequate to predict their harmful effects [2, 28].

The applications of nanotechnology are very broad, and continued evaluation of potential health hazards as the result of exposure to nanomaterials is important to ensure their safe handling. Some studies have shown that the physicochemical properties of nanomaterials can have harmful effects in the biological system [2, 22]. Initial studies of animals and humans exposed to engineered nanomaterials can provide a platform for ascertaining their harmful effects on human health and the environment. Some research studies using rodents and cell cultures have shown that the toxicity of nanomaterials is relatively higher than that of their bulk counterparts. Nanomaterials have a high surface area, high aspect ratio, high porosity, and quantum size effects. These factors have a great influence on the properties of nanomaterials. Researchers are currently engaged in nanotechnology research, but more studies are needed in order to estimate the harmful effects of nanomaterials on the human biological system. Available data on the toxicity of large particles can provide fundamental knowledge for estimating the possible adverse effects that may occur from exposure to nanomaterials. However, this information is preliminary and may not be enough to provide proper protection against nanomaterials. NIOSH has published the following summary as part of their program to minimize workplace exposure to nanomaterials [2, 22].

- Nanomaterials have the potential to enter the human body through the respiratory system if they are airborne, come in contact with the skin, or are ingested. Studies

on humans and animals have shown that airborne nanoparticles can be inhaled and deposited in the respiratory tract. Nanomaterials can enter the bloodstream and translocate to other organs in the body.
– Studies on rats show that nanoparticles are more formidable than their bulk counterparts of the same composition and may cause pulmonary inflammation and lung tumors.
– Studies in animals, cell cultures, and cell-free systems have indicated that changes in chemical composition, size of particles, and crystal structure can significantly affect oxidant generation capacity and cytotoxicity.
– Studies in workers exposed to aerosols have shown harmful lung effects, including lung function disorder and fibrotic lung disease.
– More research is needed on the adverse effects of nanomaterials.

3.6.1 Potential Safety Issues

To date, research on nanomaterials relative to their risk of fire and explosion is inconclusive. Some studies have shown that nanomaterials pose catalytic effects and explosion hazards because of their nanosize and exotic properties. Powders from nanomaterials pose a higher risk of fire and explosion than bulk-size materials of the same chemical composition because of their uncommon properties. When the material size is reduced to the nanoscale, electrons lose their freedom, which results in discrete energy states. The energy of electrons is not enough to break this confinement and, as a result, abnormal properties are observed. Decreasing the particle size of combustible materials to the nanoscale can increase the chances of spontaneous and quick combustion and cause a higher combustion rate. Some nanomaterials may initiate catalytic reactions because of their chemical composition and structure. Nanomaterials and nanostructured porous materials are used as catalysts for increasing the rate of reactions [2, 22].

3.6.2 Exposure Assessment and Characterization

It is still unknown what techniques should be used to measure and monitor the mechanisms underlying the toxicity of nanomaterials and what methods should be used to accurately predict exposure to nanomaterials in the workplace. Recent studies have shown that bulk chemistry is less important than size, shape, surface area, and surface chemistry for some nanomaterials. Techniques used for measuring nanoscale aerosols vary in complexity, but are capable of providing some useful information for evaluating occupational exposure to nanomaterials with respect to size, shape, morphology, surface area, composition, and concentration. It is very important to conduct background nanoscale particle measurements before production and processing. Personal sampling should be used to ensure representation of the exposure of workers to nanomaterials. However, a real-time exposure measurement is more helpful in evaluating the need for control systems and work practices [2, 22].

3.6.3 Precautionary Measures

Because the information currently available to predict the hazardous effects of nanomaterials is inadequate, taking measures to protect workers is highly prudent. The control of airborne exposure to nanoscale aerosols can be achieved by using a wide variety of engineering control systems, similar to those used to minimize exposure to traditional aerosols [2, 22]. The following points should be taken into account:

– Studies have shown that particle size is the key factor in causing hazardous effects in humans. Nanoparticles deposited in the human body can move quickly into organs. Research in nanotechnology is in the preliminary stages; therefore, no international standards are available to dictate what kind of clothing, gloves, and eye protection are best suited for handling nanomaterials.
– Seminars, workshops, and risk management programs must be organized in workplaces where workers are exposed to nanomaterials. These workshops and seminars should include the followings aspects:
    – Evaluation of the hazardous effects posed by nanomaterials based on their physical, chemical, and biological properties.
    – Evaluation of workforce jobs related to the handling/fabrication of nanomaterials in order to determine the potential for exposure.
    – Training and education of workers to use nanomaterials properly.
    – Installation of exhaust ventilation in areas where exposure to nanomaterials exists.
    – Compliance with general safety rules and regulations in the workplace.
    – Provision of appropriate personal protective equipment to workers.
    – Systematic evaluation of all control systems to ensure they are working properly.
    – Evaluation of the sources of error in handling nanomaterials.
– Control techniques, such as source enclosure and exhaust ventilation systems can help in capturing airborne nanoparticles. Well-designed ventilation systems with high-efficiency particulate air (HEPA) filters are now available for removing nanoparticles.
– Good work practice can help to minimize the potential hazard of nanomaterials. This includes using HEPA filters, washing hands, changing clothes daily, and preventing food consumption and beverages in places where nanomaterials are fabricated and handled.
– Isolating the sources of nanomaterials, including nanofiber and nanoparticle fabrication, from workers must be done in a closed environment or inside a fume hood. A high-efficiency ventilation system with nanoscale filters can successfully remove

nanomaterials. It is generally recommended that all nanoparticles be handled in a fume hood or in an enclosed environment.
- Respirators may be necessary when conventional control cannot adequately control exposure to nanomaterials. A specially designed and NIOSH-certified respirator can be useful for protecting workers from inhalation of nanomaterials.

3.7 Conclusions

Nanomanufacturing involves the synthesis of nanostructured materials, their impact on human health and the environment, and a life cycle assessment of nanostructured materials. The development of new synthesis techniques has invigorated the rapid emergence of a wide variety of nanoscale consumer products. Engineered nanomaterials are being used in almost all industries, and their usage is likely to increase for a wide variety of products. Such rapid advancement of nanomaterials has initiated the need for a safety assessment with respect to both human health and the environment. The potential advantages of nanomaterials are numerous, and industries have great expectations, but there are many unanswered questions regarding the safety of workers and potential hazards to humans and the environment. The risks associated with nanomanufacturing, and the use of some nanomaterials are still unclear. However, scientists are gathering data and relevant information in order to establish a database for assessing the risks and harmful effects of the synthesis and use of nanomaterials. Life cycle assessment is a comprehensive approach for documenting the impact of nanomaterial synthesis, processing, and production, using a well-defined and documented methodology. Until the hazardous effects of nanomaterials are known, precautionary measures must be taken, such as employing an advanced control system and using good work practices.

Bibliography

[1]   Genaidy A, Karwowski W. Nanotechnology occupational and environmental health and safety: Education and research needs for an emerging interdisciplinary field of study. Human Factors and Ergonomics in Manufacturing, 2006, 16(3), 247–253.
[2]   Branche CM. Approaches to safe nanotechnology, managing the health and safety concerns associated with engineered nanomaterials, Department of Health and Human Services Centers for Disease Control and Prevention, National Institute for Occupational Safety and Health, Cincinnati. DHHS (NIOSH) Publication No. 2009-125, 2009.
[3]   Aitken RJ, Chaudhry MQ, Boxall ABA, Manufacture HM. and use of nanomaterials: Current status in the UK and global trends. Occupational Medicine, 2006, 56, 300–306.
[4]   Uskokovic V. Nanotechnologies: What we do not know. Technology in Society, 2007, 29, 43–61.
[5]   Stander L, Theodore L. Environmental implications of nanotechnology – An update. International Journal of Environmental Research and Public Health, 2011, 8, 470–479.

[6]   Xia T, Li N, Nel AE. Potential health impact of nanoparticles. Annual Review of Public Health, 2009, 30, 137–150.
[7]   Meyer D, Curran MN, Gonzalez MA. An examination of existing data for the industrial manufacture and use of nanocomponents and their role in the life cycle impact of nanoproducts. Environmental Science & Technology, 2009, 43(5), 1256–1263.
[8]   Ansari A, Alhoshan M, Alsalhi MS, Aldwayyan AS. Prospects of nanotechnology in clinical immunodiagnostics. Sensors, 2010, 10, 6535–6581.
[9]   Asmatulu R, Asmatulu E, Yourdkhani A. Toxicity of nanomaterials and recent developments in the protection methods, SAMPE Fall Technical Conference, Wichita, KS, October 19–22, 2009.
[10] Aitken RJ, Creely KS, Tran CL. Nanoparticles: An occupational hygiene review. HSE Research Report, 2004, 274, London.
[11] Khan WS, Asmatulu R, Ahmed I, Ravigururajan TS. Thermal conductivities of electrospun PAN and PVP nanocomposite fibers incorporated with MWCNTs and NiZn ferrite nanoparticles. International Journal of Thermal Sciences, 2013, 71, 74–79.
[12] Tuominen M, Schultz E. Environmental aspects related to nanomaterials: A literature survey, Finnish Environment Institute, Helsinki, 2010.
[13] Patel HA, Somani RS, Bajaj HC, Jasra RV. Nanoclays for polymer nanocomposites, paints, inks, greases and cosmetics formulations, drug delivery vehicle and waste water treatment. Bulletin of Materials Science, 2006, 29(2), 133–145.
[14] Ucciferri N, Collnot EM, Gaiser KB, Tirella A, Stone V, Domenici C, Lehr MC, Ahluwalia A. In vitro toxicological screening of nanoparticles on primary human endothelial cells and the role of flow in modulating cell response. Nanotoxicology, 2014, 8(6), 697–708.
[15] Stefano B. Exposure to engineered nanomaterials and occupational health and safety effects, INAIL, Department of Occupational Health, Rome, 2011.
[16] Shrand AM, Dai L, Shlager JJ, Hussain SM. Toxicity testing of nanomaterials. In: Balls M, Combes RD, Bhogal N (eds), New technologies for toxicity testing, Springer, New York, 2012.
[17] Guang J, Haifang W, Lei Y, Xian W, Rongjuan P, Tao Y, Yulian Z, Andxinbiao G. Cytotoxicity of carbon nanomaterials: single-wall nanotube, multi-wall nanotube, and fullerene. Environmental Science & Technology, 2005, 39(5), 1378–1383.
[18] Roche Applied Science. Cytotoxicity detection kit (LDH), a non-radioactive alternative to the [3 H]-thymidine release assay and the [51Cr]-release assay, Cat. No. 11644793001, 2005.
[19] Clinton J, David GW. In vitro assessments of nanomaterial toxicity. Advanced Drug Delivery Reviews, 2009, 61(6), 438–456.
[20] Sartor M. Dynamic light scattering, University of California, San Diego, Technical Report, 2014.
[21] Chen X, Levine L, Kwok PY. Fluorescence polarization in homogeneous nucleic acid analysis. Genome Research, 1999, 9, 492–498.
[22] Asmatulu R. Nanotechnology safety, Elsevier, Amsterdam, 2013.
[23] Schrand AM, Rahman MF, Hussain SM, Schlager JJ, Smith DA, Syed AF. Metal-based nanoparticles and their toxicity assessment. Nanomedicine and Nanobiotechnology, 2010, 2(5), 544–568.
[24] Wijnhoven SWP, Peijnenburf WJGM, Herberts CA, Hagens WI, Oomen AG, Heugens EHW, Roszek B, Bisschops J, Gosens I, Meents DVD, Dekkers S, Jong WHD, Zijverden MV, Sips AJAM, Geertsma RE. Nano silver: A review of available data and knowledge gaps in human and environmental risk assessment. Nanotoxicity, 2009, 3(2), 109–138.
[25] Janer G, Mas del Molino E, Fernández-Rosas E, Fernández A, Vázquez-Compos Z. Cell uptake and oral absorption of titanium dioxide nanoparticles. Toxicity Letters, 2014, 228, 103–110.
[26] Asmatulu E, Twomey J, Overcash M. Life cycle and nano-products: End-of-life assessment. Journal of Nanoparticles Research, 2012, 14, 720. https://doi.org/10.1007/s11051-012-0720-0.
[27] Asmatulu R, Fakhari A, Wamocha HL, Hamdeh HH, Ho JC. Fabrication of magnetic nanocomposite spheres for targeted drug delivery. 2008 ASME International Mechanical Engineering Congress and Exposition, October 31–November 6, 2008, Boston, 4 pages, 2008.
[28] Nuraje N, Asmatulu R, Mul G. Green photo-active nanomaterials: sustainable energy and environmental remediation. RSC, Cambridge, England, 2015.

Jitendra S. Tate and Roger A. Hernandez

4 Safety Approaches to Handling Engineered Nanomaterials

4.1 Introduction

According to the ASTM standard E2456, nanoparticles are a subclassification of ultra-fine particles with two or three dimensions ranging from 1 to 100 nanometers [1]. Engineered nanomaterials can be defined as an engineered structures having at least one external dimension between 1 and 100 nanometers. Nanomaterials are used in many industries for products, including electronics, sporting goods, medicine, clothing, and more [2] and incorporate nanoparticles. The research and development of nanomaterials has surpassed the knowledge of their health and environmental effects. This has resulted in a need for additional safe handling guidelines, regulations, and risk assessment procedures for workers in research laboratories and manufacturing facilities that work with and are exposed to nanomaterials [3]. This article aims to provide information and guidelines for workers who use nanomaterials in an occupational setting in the hope that they follow the suggestion of improving worker safety, producing safer products, and providing insight into the steps being taken to minimize risks and impose regulatory standards.

4.2 Potential Health Concerns

There has been a worldwide outcry for a halt to nanomaterial research and production until procedures are set in place to guarantee worker safety [2]. As a particle is reduced in size, the way it interacts with its environment can change. However, due to the atypical physiochemical properties of engineered nanomaterials, limited information is available on the effects these particles can have on biological cells, especially in long-term exposure applications [4].

Cytotoxicity, the ability to be toxic to cells, is commonly referred to when discussing the potential health concerns related to nanomaterials. Properties such as surface chemistry, aspect ratio, and agglomeration state can all have diverse and adverse effects on living cells depending on the type of nanomaterial and degree of exposure [2, 5]. Epidemiological, *in vivo*, and *in vitro* studies are ways to research and collect data on nanoparticle cytotoxicity. Studies performed by the Institute of Condensed Matter Physics in Switzerland concluded that, in general, carbon-based nanomaterials, which are the most common types used industrially, lead to cell proliferation inhibition and, eventually, cell death [6]. An *in vitro* study from the Department of Metallurgical and

https://doi.org/10.1515/9783110781830-004

Materials Engineering at the University of Texas at El Paso found that a combination of carbon nanotube aggregates and carbon black was just as cytotoxic as asbestos. In contrast, silver aggregates were found to be even more cytotoxic [7, 4].

However, many engineered nanomaterials products are being produced before the hazards are fully assessed. This is problematic for the workers who manufacture these engineered nanomaterial products and end consumers [4]. Skin products, such as creams and sunblock use titanium dioxide (TiO_2) as an ultraviolet ray-blocking agent. In its bulk form, TiO_2 is considered to be harmless. Still, information on the long-term effects at the nanoscale is lacking or inconsistent [4, 8]. A number of studies on the cytotoxic effects of silicone oxide (SiO_2) have shown that it can cause cell death via apoptosis, but could not agree on the direct cause of cell death [9–11]. Other studies focusing on multi-wall carbon nanotubes (MWCNT) saw similar results, where cell damage or death was evident. Still, exact cause was not unanimous [12, 13], while one study even reported that nanotubes were not toxic [14].

Inconsistent results like these reveal the gaps of information related to nanomaterials and further demonstrate the need for standardized methods for handling nanomaterials in an occupational environment to ensure worker safety.

4.3 Proactive Measures to Examine Precautions

Before a standard set of regulations can be put into effect, it is important to know the practices currently in use throughout different industries to better understand where improvements in nanomaterial safety knowledge should be made [3].

Many surveys have been conducted over the years, examining how different companies handle and assess nanomaterials. One survey focusing on nanomaterial risk and safety issues in German and Swiss industries asked 40 companies a series of questions about the kinds of nanomaterials they use in their products and how, if at all, they assess these hazards. Some of these questions included: "What are the mean particle diameter and the particle size distribution of the nanoparticulate material (NPM) in your product?", "Have you evaluated the possible uptake of the NPM by the following organisms (aquatic, soil, humans, other organisms) during the different stages of the product life cycle?", and "Does your company conduct risk assessments where NPM are involved?". The results concluded that 26 of the 40 companies surveyed had no nanomaterial risk assessment procedures [15].

A second survey, published in 2012, asked 78 companies in 14 countries about how they perceive the risks related to different types of nanomaterials (carbon nanotubes, quantum dots, metal oxides, heavy metals, dry powders, and other carbonaceous materials). The study found that 22 % and 40 % of participants admitted that they do not know the risks of CNTs and quantum dots, respectively. An average of 28 % of companies believed there was little to no risk involved with any of the nanomaterials described in the survey. In comparison, 44 % reported moderate to high risks associated with each nanomaterial. On average, almost one-third of participants reported no information on the risks of any nanomaterials [16].

4.4 Assessment of Engineered Nanomaterials

As with any material, it is good practice for a worker to know what kind of material they are working with and how to handle it. If there is little or no knowledge available about the material, then the material should be considered hazardous and treated as such. To prevent a lack of knowledge, it is highly recommended to preemptively collect as much useful information on a material as possible. For example, suppliers may provide a safety data sheet, which may include some, but not all, of the required information about a material. This information, however, may only represent the material on the macroscale, not the nanoscale, as properties can change when particle size is reduced. In this event, additional research should be conducted to protect the workers and anyone who may come in contact with the material at any point during its life cycle. This information includes but is not limited to [17–19]:

–   Commercial and technical names
–   Current safety data sheets (SDS) and technical data sheets (TDS)
–   Chemical composition
–   Presence of nanomaterials and identification
–   Proportions of nanomaterials
–   Particle size distribution
–   Material dustiness
–   Solubility
–   Hazard and toxicity levels
–   Material substitutions

4.4.1 Hazard Assessment

Nanoparticles can exhibit changes in their physical, chemical, and mechanical properties compared to the same material in bulk form. Two key reasons for this alteration are surface effects and quantum effects [4]. Surface effects cause the smooth scaling of properties due to the fraction of atoms at the surface. In contrast, quantum effects display discontinuous behavior due to quantum confinement effects in materials with delocalized electrons [20]. It is important to know which properties of a material can change and the effects it can cause when one is to be working in an environment containing nanomaterials.

Studies on the toxicity of single-wall carbon nanotubes (SWCNT) on mice have shown that the physical and chemical properties of nanoparticles can cause serious health issues [21, 22]. These properties include [23, 18]:
– Particle size and distribution
– Nanomaterial shape
– Agglomeration state
– Solubility
– Surface area
– Porosity
– Reactivity
– Impurities and/or contaminants
– Chemical composition
– Physical properties
– Crystal structure

4.4.2 Hazardous Communication

It is the responsibility of the employer to communicate information to employees who may come in contact with engineered nanomaterials in regards to any and all possibilities of hazardous exposure. This list, which has been derived from the ASTM standard guide for "Handling Unbound Engineered Nanoscale Particles in Occupational Settings," explains the type of information employers must communicate to workers with respect to either normal or emergency conditions [23].
– Any known and potential physical, health, and safety hazards related to engineered nanomaterials
– Which processes in the work environment come in contact with engineered nanomaterials
– How to determine the presence of engineered nanomaterials in the work environment by visual appearance, odor, etc.
– Procedures on exposure minimization, including engineering and administrative controls, personal protective equipment, and emergency procedures

4.4.3 Exposure Assessment

So far, no general occupational exposure limit is primarily concerned with airborne exposure of unbound engineered nanoparticles. There are, however, designated occupational exposure limits for nuisance (poorly soluble or insoluble) particles that can be used as a reference for particles of similar physical and/or chemical composition. However, these limits do not necessarily take particle size into account and may be inaccurate for nanoscale nuisance particles [24].

When assessing exposure, it is crucial to understand how nanoparticles can enter the body. There are five types of exposure routes:
- Inhalation
- Ingestion
- Dermal
- Ocular
- Injection

Inhalation, ingestion, and dermal are the three most common and most discussed methods of exposure, as they typically deal with airborne particles: inhalation being the most common and most studied of the three [25].

Engineered nanoparticles bound in a solid matrix, such as nanocomposites, pose little to no exposure risk while being handled. There may be a possibility of exposure if solid engineered nanomaterials are machined or burnt, unbinding the nanoparticles from the matrix in the form of airborne dust. This can lead to inhalation, ingestion, and dermal or ocular exposure. Nanoparticles bound in a liquid matrix, such as resins, pose a higher risk of exposure. Physical contact with nanoparticles suspended in a liquid matrix can result in dermal exposure, while inhalation and/or ingestion exposure is possible if the particles become aerosolized [23].

When working with nanoparticles, it is best to use these manipulation methods in this order from least to most favorable as displayed in Figure 4.1 [17, 24, 19].

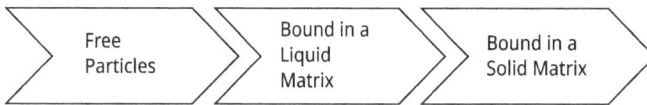

Figure 4.1: Nanoparticle manipulation method precedence.

In order to properly assess exposure, a facility should keep records of any processes where engineered nanomaterials are present [26]. These processes can include:
- Material receiving/shipping
- Manufacturing processes
- Material handling
- Storage
- Waste management
- Maintenance
- All other activities performed during an ENM's lifecycle

4.5 Characterization of Engineered Nanomaterials

Characterization can be defined as determining the physical and chemical properties of a material. These can include size, shape, atomic structure, and solubility, among many others [27]. According to the standard ISO/TS 12901-1, some of the most commonly used engineered nanomaterials include fullerenes, carbon nanotubes, carbon black, quantum dots, and metals and metal oxides [17].

4.5.1 Fullerenes

Fullerenes are comprised completely of carbon. They take on a hollow sphere shape and form hexagonal carbon ring structures, similar to graphite. However, pentagonal and heptagonal rings are possible, allowing for three-dimensional structures. A common form of fullerene is the Buckminster fullerene (buckyball). When under the right conditions, they can behave as a semiconductor, a conductor, or a superconductor. Fullerenes have the ability to contain other materials and substances inside and also absorb free radicals. They are also chemically inert and relatively harmless to humans [17].

4.5.2 Carbon Nanotubes

Carbon nanotubes, a type of fullerene, are another commonly used nanomaterial and can be classified into three categories: Single-walled carbon nanotubes, Double-walled carbon nanotubes, and Multi-walled carbon nanotubes [17]. SWCNT consist of sheets of hexagonal ring structured graphene sheets rolled into a cylinder along a lattice vector. CNT can be relatively long, ranging from 1–100 μm, with 1–2 nm diameters for SWCNT and over 100 nm for MWCNT. This high aspect ratio results in excellent electrical conductivity and can be applied to plastic and composite materials. This conductivity can be increased or decreased based on the CNT's diameter and chirality, the degree between hexagons and the tube's axis [18, 28]. According to a study on the cytotoxicity of CNT on human cells, as a result of their surface area and surface chemistry, refined SWCNT caused the most severe cell damage compared to other types of nanomaterials tested. Even a minuscule amount of refined SWCNT (25 μg/ml) increased cellular apoptosis/necrosis compared to 0.06 mg/ml of unrefined SWCNT. The difference between refined and unrefined SWCNT is that refined SWCNT do not possess catalytic transition metals, such as iron or cobalt [21, 22].

4.5.3 Carbon Black

A common nanomaterial used in the industrial setting is carbon black. Carbon black, known for its electrical conductivity, relatively high surface area to volume ratio, high tinting strength, and ability to absorb ultraviolet light, can be classified into five subcategories: furnace black, thermal black, acetylene black, channel black, and lamp black. Those categories can be further dissected based on particle grade and size [29, 30]. Generally, carbon black contains more than 97 % elemental carbon arranged in an aciniform structure (grape cluster). A common practice employing carbon black's electrical conductivity is using it as filler in plastic and elastomer products as an antistatic agent. Almost 90 % of carbon black is used as filler in rubber products, most notably in tires. 9 % of carbon black is used as a pigment in paints, inks, and resins, including newspapers, inkjet printers, and automobiles. The other 1 % is utilized in countless applications throughout various industries [31, 32]. According to a study on high-priority carcinogens, carbon black may be carcinogenic to humans. This conclusion came after a series of animal trials and a few tests conducted on humans [33, 30].

4.5.4 Quantum Dots

Quantum dots, with their core-shell structure, are composed of semiconductor nanocrystals and range in diameter from 2 to 10 nanometers [18, 17]. The core and shell of a quantum dot are usually made up of two different combinations of elements. The core typically comprises two elements from groups 2B and 6A, 3A and 5A, or 4A and 6A. The shell is usually made of zinc sulfide [34]. Semiconductor quantum dots possess photophysical properties related to their size. During excitation, a desired wavelength of light can be achieved by altering the particle size. This is especially useful for *in vitro* diagnostics, therapeutics, imaging, and optical devices [35]. However, the use of quantum dots is limited due to cytotoxicity concerns surrounding the nature of their semiconductor elements and outer chemical coatings succumbing to oxidation or U.V. damage in biological systems [36].

4.5.5 Metals and Metal Oxides

A wide category of nanoparticles includes metals and metal oxides. There is no set definition on the size and shape of these nanoparticles, as they can consist of several different types of metals and alloys [17]. Commonly used metal nanoparticles are titanium dioxide, gold, and silver. These particles' small sizes typically change the material's properties compared to their macroscopic counterparts. For example, the color of gold nanoparticles changes as the particle size increases or decreases [37]. Although it has not been concluded that metal oxide nanoparticles are always more toxic than metal oxide microparticles, copper oxide nanoparticles have been found to be highly toxic compared to copper oxide microparticles [24]. Titanium dioxide (TiO_2), a commonly used component in sunscreen, is categorized as a possible carcinogen based on sufficient studies on animals and an inadequate number of studies on humans [30]. In light of this, cytotoxicity is still a concern with metal nanoparticles, but most applications use these nanoparticles in sizes too large for dermal penetration. Typical applications of metals and metal oxides can include cosmetics, coating, pigments, and composites [38, 18].

4.6 Control Preferences

When handling nanoparticles, nanomaterials, and hazardous materials, it is important to understand how to minimize worker exposure. Figure 4.2 shows a standardized hierarchy of exposure control that should be implemented to ensure worker safety. In addition, trt displays different exposure control methods ranked from first method of action (top) to last resort (bottom) [17, 24, 19].

1. Elimination: removing any hazardous materials entirely from a process or workplace
2. Substitution: replacing hazardous materials with ones that pose less risk to workers
3. Isolation: isolating a process or machine in an area where hazardous materials can be used without causing harm to workers
4. Engineering Controls: altering a process or work environment to reduce exposure
5. Administrative Controls: implementing safety precautions and good housekeeping
6. Personal Protective Equipment: equipment worn by workers to protect from hazardous materials

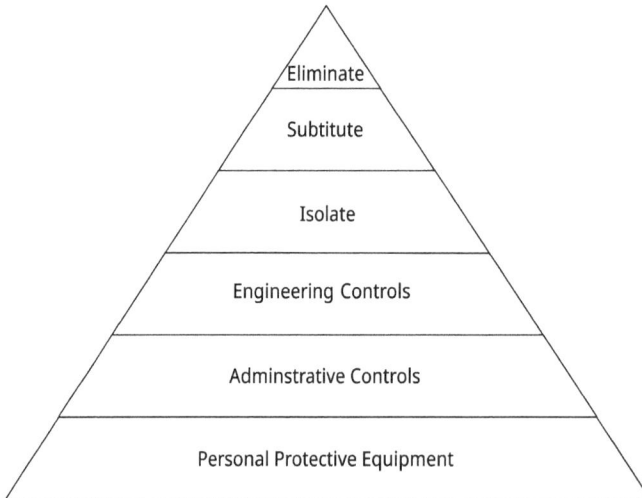

Figure 4.2: Exposure Control Hierarchy.

1. Elimination: If a hazardous material is not necessary, it should be avoided whenever possible. Completely removing hazardous materials from a process may not always be an option. Substitution is the next step [26, 24].

2. Substitution: Replacing hazardous materials, like toxic chemicals, nanoparticles, etc., with less hazardous ones is an excellent way to protect workers from exposure. This only works, however, if the new, less hazardous materials can perform the same functions as the previous materials. If not, isolation may be necessary [26, 24].

3. Isolation: In situations where there is no other choice but to use a hazardous material in a process, isolating that process should be considered. An example of this would be placing a machine, such as a sonicator used for nanoparticles, into an isolated room, where particles and exhaust can be contained, and workers outside are free from exposure [26, 24].

4. Engineering Controls: To reduce the amount of exposure, changes in a process or the workplace may be required. These changes include containing a process in a separate environment (isolation), and ventilation [39].
– General Exhaust Ventilation
    – General ventilation for a workplace
    – Usually provided by an HVAC system
    – Should be placed as close to the source of contamination as possible
    – Exhaust should not contaminate any areas of air intake or be recirculated back into the work environment
– Local Exhaust Ventilation
    – Contains contaminants in a small area or completely encloses them
    – Includes chemical fume hoods, ventilated enclosures, and snorkel hoods
    – Exhaust is funneled into an air cleaner or filtered before being released into the open air
    – HEPA (high-efficiency particulate air) filters should be used on any exhaust ventilation where engineered nanomaterials are to be handled

These ventilation systems should employ negative pressure to ensure adequate protection. For example, g boxes and other sealed enclosures offer a higher level of protection compared to other, less pressurized enclosures, such as ventilated hoods and snorkel hoods [26, 24].

5. Administrative Controls: Worker safety does not just fall back on the employee but also on the employer. It is the responsibility of both the employee and employer to work in tandem to ensure proper safety conditions are met and maintained in the work environment. Administrative controls are generally used to limit worker exposure to hazardous materials and promote worker safety [23]. Proper hand washing facilities should be provided to workers along with basic hygienic information: washing hands after coming into contact with engineered nanomaterials and before consuming food or drink [40]. Also, employers should reduce how long workers are exposed to engineered nanomaterials by regulating working hours and implementing shifts to rotate workers.

Typically, general workplace administrative controls can be applied to an environment containing engineered nanomaterials, provided the additional precautionary measures and practices required for such environments are also implemented.

- *Housekeeping*: General good housekeeping should be practiced in any workplace. Keep work areas clean and clear of debris, regularly clean work surfaces, replace any tools or equipment to their original storage location, etc. In the case of a work environment where engineered nanomaterials are handled, ensure all equipment, including glassware, tools, and ventilation equipment, are vacuumed with a HEPA filter and/or cleaned and wiped regularly (minimum at the end of each shift). Dry sweeping and using compressed air to clean up nanoparticle dust should be prohibited. These procedures can also fall under work practices [39, 24].
- *Work Practices*: Another step in promoting worker safety is employing good work practices. To do so, employers and employees must understand the potential hazards of engineered nanomaterials and other hazardous materials. In addition, OSHA has prepared a list of work practices (Table 4.1), that pertains to both management and workers to ensure proper worker safety. In addition, administrative controls can fall under the category of work practices, such as housekeeping and training [28, 26].
- *Signs/Labels/Storage*: Areas, where engineered nanomaterials are present, should be posted with a hazard sign stating as such. This includes isolated work areas, exhaust ventilation areas, and any other workplace where engineered nanomaterials are handled. Signs should also indicate what personal protective equipment

Table 4.1: Good work practices for management and workers according to OSHA.

Management	Workers
Provide education to workers on safe handling of nanomaterials to reduce the possibility of exposure.	Avoid working with nanomaterials in a free particle state in open air conditions.
Provide information to workers on any hazardous properties of materials being handled and how to prevent exposure.	Store dispersible nanomaterials in tightly sealed containers when possible. (Both liquid suspended and dry particles)
Encourage workers to use sanitary facilities to wash their hands before leaving the workplace, smoking, or consuming food or drink.	Clean work areas and equipment at the end of each shift, at the minimum. Avoid dry sweeping or compressed air. Vacuum using a HEPA filter, or use wet wiping methods.
Provide additional control measures to ensure nanomaterials do not leave designated work areas. (Decontamination facilities, buffer areas, etc.)	Waste material must be disposed of in compliance with any and all federal, state, and local regulations.
Provide showering and changing facilities for workers to prevent contamination of other areas caused by clothing and skin transfer of nanomaterials.	Food and beverages should not be stored or consumed in areas where nanomaterials are handled or stored.

is required to enter the area. Storage areas for engineered nanomaterials should also have proper signage to indicate hazardous materials. According to the "OSHA Hazard Communication Standard, 29 CRF 1910.1200," employers must label all hazardous chemicals and materials in the work environment [24]. Appropriate labels must indicate contents and material form (liquid, powder, aerosol, nanoparticulate, etc.) when storing engineered nanomaterials. Contact information for a facility's Environmental Health and Safety Office, or other appropriate representative, should be provided on the label in case of container breakage. Secondary containers must also be labeled appropriately [39].

- *Training*: Any worker required to handle engineered nanomaterials must be trained on all hazards and risks. OSHA requires that, at a minimum, workers be trained on detecting chemicals in the work environment, any hazards associated with the chemicals, and methods to avoid exposure [40]. Additional training should incorporate explanations of relevant safety data sheets and labeling systems, proper engineered nanomaterial handling and storage procedures, appropriate personal protective equipment usage, proper cleaning methods of engineered nanomaterial contaminated surfaces, and proper engineered nanomaterial waste disposal [23, 17].

6. Personal Protective Equipment: When engineering and administrative controls cannot provide a safe work environment from exposure, the final step is to employ personal protective equipment. Employers must provide the necessary PPE to the appropriate workers and provide education and training on how to use it properly. Figure 4.3 lists different types of PPE along with appropriate annotations for proper usage, disposal, etc. [41, 26].

4.7 Management of Engineered Nanomaterials

Engineered nanomaterials should be always be considered hazardous. For this reason, waste disposal, spills, and releases should be strictly regulated. Most regulations typically refer to any engineered nanomaterials that are not bound in a solid matrix, or bound in a solid matrix, but present the risk of breaking free when contacted with air or water. This includes engineered nanomaterials bound in a liquid matrix, PPE, and cleaning wipes contaminated with engineered nanomaterials. In addition, disposal of engineered nanomaterials should follow any existing federal, state, or local regulations where the facility resides [39].

4.7.1 Waste Disposal

As with any hazardous material, engineered nanomaterials should have their designated disposal area. Engineered nanomaterials should not, under any circumstances,

Pants	• Long pants • No cuffs (can catch airborne nanomaterial)
Shoes	• Closed-toed shoes • Low permeable material • Disposable booties may be worn over shoes to prevent tracking of nanomaterials out of work area
Lab Coat (disposable)	• Impervious materials (noncotton) • Should be treated as hazardous waste
Lab Coat (non-disposable)	• Should remain in work area • Place in sealed bags before removing from work area to be cleaned • Inform cleaning service of contamination
Gloves	• Latex or nitrile gloves should be worn when handling nanomaterials • Change gloves frequently • Store used contaminated gloves in a plastic bag until properly disposed of as hazardous waste • More porous gloves, such as cotton, may be worn when handling solid bound nanomaterials
Eye Protection	• Select eye protection based on hazard level • Safety glasses with side shields offer minimum protection • Full face masks or helmet respirators require no further eye protection
Respirator	• Workers must receive medical clearance by a medical professional before being fitted • Wear appropriate respirator and cartridge based on hazard level • A half mask respirator should be worn at a minimum • Refer to your EHS office to confirm proper respirator use for your facility

Figure 4.3: Proper PPE for a nanomaterial contaminated work environment.

be placed in the regular trash or poured down the drain. Any engineered nanomaterial waste should be properly stored in a small sealed container with appropriate labels indicating the type of waste, the form it is in, and any known properties related to the material. The label should also explicitly state that the container is holding nanomaterials [39, 23, 26]. Larger containers are not preferred, as container failures are much more challenging to clean up than smaller ones. This is especially true with more dispersible forms of nanomaterial waste [17]. Any other loose forms of contaminated waste, such as PPE, wipes, gloves, etc., should be placed in a sealable container or resealable plastic bag and stored in a fume hood or other well-ventilated area. Once the container or bag is full, double-bag it, attach the appropriate labels, and ensure it is appropriately sealed before removing it from the fume hood. Waste containers should then be properly disposed of according to the facility's standard procedures [28].

4.7.2 Management of Spills

In the event of a nanomaterial spill, spill kits should be available. According to OSHA, spill kits should contain at the minimum [24]:
– Barricade tape
– Nitrile or latex gloves
– Elastomeric respirator with filters

- – Adsorbent material
- – Wipes
- – Sealable plastic bags
- – Walk-off mat
- – Vacuum with HEPA filter
- – Spray bottle with deionized water

The type of spill warrants the type of procedure needed for cleanup. When assessing a spill, determine how large of an area is contaminated. Block off this area using barricade tape or other means of entry restriction. If the spill is minor (less than a few grams of nanomaterial), it may be cleaned by trained personnel in the work area. If the spill is more significant (more than a few grams of nanomaterial), then restrict any entry to the area and contact the Environmental Health and Safety Office for further instructions [18, 26].

For liquid spills, such as nanomaterials suspended in a liquid matrix, standard hazardous material cleanup procedures should be followed. These will be based on the type of material and its known hazards. Additionally, extra steps should be taken to remove any nanomaterials altogether. Barriers should be positioned around the area to minimize air currents moving across the spill. An absorbent walk-off mat should be placed at the exit point of the contaminated area to prevent nanomaterials from being tracked out. Vacuum up any excess nanomaterial with a HEPA-filtered vacuum. Once the area is clear, treat any contaminated cleaning materials as hazardous waste and dispose of them properly [39, 28].

Similar procedures to liquid spills can be followed for dry materials, like powders. A sticky walk-off mat should be placed at the exit point of the contaminated area. Use wet wipes to clean small dry spills, then dispose of the wipes as hazardous waste. For larger dry spills, contact the Environmental Health and Safety Office. If given the OK, vacuum the spill using a HEPA-filtered vacuum cleaner; otherwise, wait for further instructions. Do not dry sweep nanomaterials, as this can cause airborne dispersion. This includes broom and compressed air sweeping [23, 39].

4.8 Overview of National and International Associations that Adopted the Handling and Use of Nanomaterials

4.8.1 British Standards Institution (BSI)

Beginning in 2008, the British Standards Institution published guidance consisting of nine nanotechnology documents covering information regarding terminology for medical, health, and personal care applications of nanotechnologies, bio-nano interface, nanoscale measurement and instrumentation, carbon nanostructures, nanofabrication, nanomaterials, a guide for specifying manufactured nanomaterials, and a guide to safe handling and disposal of manufactured nanomaterials. Various experts have developed these nine documents from industry, academia, Government, and professional organizations, all congregated by the British Standards Institution. These documents are a part of the U.K.'s contemporary work to develop standards for a field that is still relatively new [18].

4.8.2 Health and Safety Executive (HSE)

The Health and Safety Executive (hereafter referred to as "the Executive") was founded under the Health and Safety at Work etc. Act of 1974, along with the Health and Safety Commission. As of 2007, the Health and Safety Commission has delegated its responsibilities to the Executive.

The duties of the Executive, granted by the Safety at Work, etc. Act of 1974 include the following:
a) "to assist and encourage persons concerned with matters relevant to those purposes to further those purposes;"
b) "make such arrangements as it considers appropriate for the carrying out of research and the publication of the results of research and the provision of training and information, and encourage research and the provision of training and information by others;"
c) "make such arrangements as it considers appropriate to secure that the following persons are provided with an information and advisory service on matters relevant to those purposes and are kept informed of and are adequately advised on such matters:
   i.   government departments,
   ii. local authorities,
   iii. employers,
   iv. employees,
   v.   organizations representing employers or employees, and
   vi. other persons concerned with matters relevant to the general purposes of this Part."

The Executive also proposes regulations to and advises the Secretary of State based on research and the publication of the study results on behalf of the Safety at Work Act.

The HSE endorses the U.K. NanoSafety Group. The group has published guidance for research and development and academia for managers, advisors, employers, and users of nanomaterials. The guidance covers multiple nanomaterials, such as fibers, powders, tubes, wires, and aggregates. The U.K. NanoSafety Group aims to promote a precautionary strategy to minimize the risk of nanoparticle exposure. In addition, the group cohosted an international conference on Working Safely with Nanomaterials at the Royal College of Physicians to spread nanosafety awareness further [26].

4.8.3 International Organization for Standardization (ISO)

The International Organization for Standardization (ISO) started in 1946 when delegates from 25 countries met in London to create an international organization "to facilitate the international coordination and unification of industrial standards." The organization began operation the following year. ISO is now a network of national standards bodies consisting of 162 member countries, a Central Secretariat based in Switzerland, and 150 full-time employees.

The function of ISO is to develop standards with the help of a variety of experts from all over the world. Each group of experts develops standards based on requirements in their respective sectors. ISO has over 19,500 standards touching nearly all aspects of life, such as air, water, and soil quality, emissions of gases and radiation, and environmental aspects of products. These standards contribute to efforts to preserve the environment and protect the health of people around the world [42].

In 2005, ISO began researching and developing nanotechnology and deriving standards based on those results. Standardization in the field of nanotechnologies includes 86 published ISO standards about understanding and control of matter and processes at the nanoscale and using various properties of nanomaterials that differ from the properties of individual atoms, molecules, and bulk matter, to create better materials, devices, and systems that feature the new properties. Specific tasks for the International Organization for Standardization include developing standards for the following: "terminology and nomenclature; metrology and instrumentation, including specifications for reference materials; test methodologies; modeling and simulations; and science-based health, safety, and environmental practices." Currently, 34 countries are involved in ISO's nanotechnology research [43].

4.8.4 Organization for Economic Cooperation and Development (OECD)

The Organization for European Cooperation (OEEC) was created in 1948 and was designed to run the Marshall Plan to reconstruct Europe's damaged infrastructure caused by World War II. The OEEC helped countries realize that their economies were interdependent, bringing with it a new era of cooperation that could transform Europe. Once other nations worldwide saw the OEEC's success, they realized the need for an organization like this on a world stage. In 1960, representatives from Canada and the United

States joined OEEC members in signing the new Organization for Economic Cooperation and Development Convention. OECD began operations on September 30, 1961, when the Convention entered into force. Since then, the OECD's number of member countries has grown to 34. These countries work together to identify problems, analyze them, and then promote policies to solve them. Since the OECD was created, some countries have seen their national wealth double or triple. Countries that were only minor players on the world stage, such as Brazil and India, are now key partners of the OECD and contribute to its work [44].

For over 40 years, the Organization for Economic Cooperation and Development has been pivotal in safely using chemicals and protecting human health and the environment. In response to emerging issues, the OECD launched the "Sponsorship Programme for the Testing of Manufactured Nanomaterials" in 2006 to improve hazard, exposure, and risk assessment approaches for manufactured nanomaterials. "OECD complements the testing program with guidance on exposure measurement (including sampling techniques and protocols) and exposure mitigation." These documents address exposure at the workplace, exposure to consumers, and environmental exposure. In addition, the OECD has been vital in analyzing risk assessment strategies for manufactured nanomaterials and whether current waste management practices are adequate for nanomaterial waste [45].

4.8.5 U.S. National Institute for Occupational Safety and Health

The National Institute for Occupational Safety and Health (NIOSH) was established under section 22 of the Occupational Safety and Health Act of 1970 as a part of the Centers for Disease Control and Prevention. The duties of the organization under the OSHA Act are to

a) "develop recommendations for health and safety standards."
b) "develop information on safe levels of exposure to toxic materials and harmful physical agents and substances."
c) "conduct research on new safety and health problems". NIOSH may also "conduct on-site investigations (Health Hazard Evaluations) to determine the toxicity of materials used in workplaces."
d) "fund research by other agencies or private organizations through grants, contracts, and other arrangements."

The mission of NIOSH is to produce scientific knowledge and provide practical solutions to reduce the risks of workplace injuries and deaths. NIOSH employs over 1,300 people from multiple fields, including medicine, industrial hygiene, epidemiology, psychology, chemistry, statistics, and many branches of engineering. Headquartered in Washington, D.C., and Atlanta, Georgia, NIOSH has eight laboratories nationwide to research the many emerging problems arising from dramatic changes in the 21st Century workplace and workforce [46].

In 2004, NIOSH established the NIOSH Nanotechnology Research Center (NTRC). The purpose of The NTRC is to accelerate progress in research into the application of nanoparticles and nanomaterials in occupational safety and health and the implications of nanoparticles and nanomaterials for work-related injury and illness. The NTRC consist of NIOSH scientist from various disciplines, who are responsible for developing and guiding NIOSH scientific and organizational plans in nanotechnology health research. One year after creating The NTRC, NIOSH drafted the Strategic Plan for NIOSH Nanotechnology Research: Filling the Knowledge Gaps. The plan called for a concerted effort to identify and address the knowledge gaps in nanotechnology. The four main goals for the NIOSH nanotechnology research are as follows [47]:

- "Understand and prevent work-related injuries and illnesses potentially caused by nanoparticles and nanomaterials."
- "Conduct research to prevent work-related injuries and illnesses by applying nanotechnology products."
- "Promote healthy workplaces through interventions, recommendations, and capacity building."
- "Enhance global workplace safety and health through national and international collaborations on nanotechnology."

4.8.6 Safe Work Australia (SWA)

Safe Work Australia was established by the Safe Work Australia Act 2008. It began operating as an independent Australian Government statutory agency in 2009. Safe Work Australia mainly comprises members representing the Commonwealth, the States, the Territories, workers, and employers. SWA's primary responsibility is to lead policy development to improve occupational health and safety and workers' compensation arrangements in Australia. The Key functions of Safe Work Australia as set out in the Safe Work Australia Act of 2008 are to

a. "develop national policy relating to occupational health and safety and workers' compensation."
b. "prepare a model Act and model regulations relating to occupational health and safety and, if necessary, revise them:"
c. "prepare model codes of practice relating to occupational health and safety and, if necessary, revise them:"
d. "prepare other material relating to occupational health and safety and, if necessary, revise that material."
e. "develop a policy, for approval by WRMC, dealing with the compliance and enforcement of the Australian laws that adopt the approved model occupational health and

safety legislation, to ensure that a nationally consistent approach is taken to compliance and enforcement."

f.   "monitor the adoption by the Commonwealth, states, and territories of:"

g.   "collect, analyze and publish data or other information relating to occupational health and safety and workers' compensation to inform the development or evaluation of policies about those matters."

h.   "conduct and publish research relating to occupational health and safety and workers' compensation to inform the development or evaluation of policies concerning those matters."

i.   "revise and further develop the National Occupational Health and Safety Strategy 2002–2012 released by WRMC on May 24, 2002, as amended from time to time"

j.   "develop and promote national strategies to raise awareness of occupational health and safety and workers' compensation."

k.   "develop proposals relating to:"

l.   "advise WRMC on matters relating to occupational health and safety or workers' compensation."

m.   "liaise with other countries or international organizations on matters relating to occupational health and safety or workers' compensation, and,"

n.   "perform such other functions that are conferred on it by WRMC."

In January of 2010, Safe Work Australia published the Work Health and Safety assessment tool for handling engineered nanomaterials. The assessment tool will be used to document practices and procedures and work with health and safety regulators when visiting nanotechnology organizations. The assessment tool provides a checklist of nanomaterials and how they're being used, controls in place to prevent exposure, information available to businesses, organizations, or labs, and also characteristics of the business manufacturing, supplying, or using nanotechnology. This tool plays a vital role in helping manufacturing facilities document their use of engineered nanomaterials and assisting regulators in checking if the proper precautions are being taken when handling these nanomaterials [48].

Bibliography

[1]   ASTM Standard E2456-06. Standard terminology relating to nanotechnology, ASTM International, West Conshohocken, PA, US, 2020.

[2]   Nel A, Xia T, Mädler L, Li N. Toxic potential of materials at the nanolevel. Science, 2006, 622–627.

[3]   Conti JA, Killpack K, Gerritzen G, Huang L, Mircheva M, Delmas M, Holden PA. Health and safety practices in the nanomaterials workplace: results from an international survey. Environmental Science & Technology, 2008, 3155–3162.

[4]   Pacheco-Blandino I, Vanner R, Buzea C. Toxicity of nanoparticles. In: Pacheco-Torgal F, Jalali S, Fucic A (eds). Toxicity of building materials (pp. 427–463). Woodhead Publishing Limited, Cambridge, 2012.

[5]   Wiesner MR, Lowry GV, Alavarez P, Dionysiou D, Biswas P. Assesing the risks of manufactured nanomaterials. Environmental Science & Technology, 2006, 4336–4345.

[6]   Magrez A, Kasas S, Salicio V, Pasquier N, Seo J, Celio M, Forró L. Cellular toxicity of carbon-based nanomaterials. Nano Letters, 2006, 1121–1125.

[7]   Soto KF, Carrasco A, Powell TG, Garza KM, Murr LE. Comparative in vitro cytotoxicity assessment of some manufactured nanoparticulate materials characterized by transmission electron microscopy. Journal of Nanoparticle Research, 2005, 145–169.

[8]   Soharbuddin SK, Thevenot PT, Baker D, Eaton JW, Tang L. Nanomaterial cytotoxicity is composition, size, and cell type dependent. Particle and Fibre Toxicology, 2010, 1–17.

[9]   Thibodeau MS, Giardina C, Knecht DA, Helble J, Hubbard AK. Silica-induced apoptosis in mouse alveolar macrophages is initiated by lysosomal enzyme activity. Toxicological Sciences, 2004, 34–48.

[10]   Wang L, Bowman L, Lu Y, Rojanasakul Y, Mercer RR, Castranova V, Ding M. Essential role of p53 in silica-induced apoptosis. American Journal of Physiology. Lung Cellular and Molecular Physiology, 2003, 488–496.

[11]   Fubini B, Hubbard A. Reactive oxygen species (ROS) and reactive nitrogen species (RNS) generation by silica in inflammation and fibrosis. Free Radical Biology & Medicine, 2003, 1507–1516.

[12]   Bottini M, Bruckner S, Nika K, Bottini N, Bellucci S, Magrini A, Mustellin T. Multi-walled carbon nanotubes induce T lyphocyte apoptosis. Toxicology Letters, 2006, 121–126.

[13]   Grabinski C, Hussian S, Lafdi K, Braydich-Stolle L, Schlager J. Effect of particle dimension on biocompatibility of carbon nanomaterials. Carbon, 2007, 2828–2835.

[14]   Chlopek J, Czajkowska B, Szaraniec B, Frackowiak E, Szostak K, Béguin F. In vitro studies of carbon nanotubes biocompatibility. Carbon, 2006, 1106–1111.

[15]   Helland A, Scheringer M, Siegrist M, Kastenholz HG, Wiek A, Scholz RW. Risk assessment of engineered nanomaterials: a survey of industrial approaches. Environmental Science & Technology, 2008, 640–646.

[16]   Engeman CD, Baumgartner L, Carr BM, Fish AM, Meyerhofer JD, Satterfield TA, Holden P, Harthorn BH. Governance implications of nanomaterials companies' inconsistent risk perceptions and safety practices. Journal of Nanoparticle Research, 2012, 1–12.

[17]   ISO. ISO/TS 12901-1. Nanotechnologies-occupational risk management applied to engineered nanomaterials-part 1: principles and approaches. Gevena, Switzerland: ISO, 2012. Retrieved from www.iso.org.

[18]   BSI. Nanotechnologies – part 2: guide to safe handling and disposal of manufactured nanomaterials. British Standards Institution, London, United Kingdom, 2007.

[19]   Warheit DB, Sayes CM, Kenneth RL, Swain KA. Health effects related to nanoparticle exposures: Environmental, health and safety considerations for assessing hazards and risks. Pharmacology & Therapeutics, 2008, 35–42.

[20]   Buzea C, Pacheco II, Robbie K. Nanomaterials and nanoparticles: sources and toxicity. Biointerphases, 2007, 1–56.

[21]   Tian F, Cui D, Schwarz H, Estrada GG, Kobayashi H. Cytotoxicity of single-wall carbon nanotubes on human fibroblasts. Toxicology in Vitro, 2006, 20(7), 1202–1212.

[22]   Maynard AD, Kuempel ED. Airborne nanostructured particles and occupational health. Journal of Nanoparticle Research, 2005, 587–614.

[23]   ASTM Standard E2535-07. Standard guide for handling unbound engineered nanoscale particles in occupational settings, ASTM International, West Conshohocken, PA, US, 2013. https://doi.org/10.1520/E2535.

[24]   NIOSH. General safe practices for working with engineered nanomaterials in research laboratories. Atlanta, GA, United States: United States Department of Health and Human Services: Centers for Disease Control and Prevention, National Institute for Occupational Safety and Health, DHHS (NIOSH), 2012. Retrieved from www.cdc.gov/niosh.

[25] Maynard A. Nanomaterials and Occupational Health. Cinncinnati, OH: United States Department of Health and Human Services: Centers for Disease Control and Prevention, National Institute for Occupational Safety and Health, DHHS (NIOSH), 2004.

[26] NIOSH. Current strategies for engineering controls in nanomaterial production and downstream handling processes. Cincinnati, OH: United States Department of Health and Human Services: Centers for Disease Control and Prevention, National Institute for Occupational Safety and Health, DHHS (NIOSH), 2013.

[27] Oberdörster G, Maynard A, Donaldson K, Castranova V, Fitzpatrick J, Ausman K, Yang H. Principles for characterizing the potential human health effects from exposure to nanomaterials: elements of a screening strategy. Particle and Fibre Toxicology. 2005.

[28] NIOSH. Approaches to safe nanotechnology: managing the health and safety concerns associated with engineered nanomaterials. Cincinnati, OH, Ohio, United States: United States Department of Health and Human Services: Centers for Disease Control and Prevention, National Institute for Occupational Safety and Health, DHHS (NIOSH), 2009.

[29] Ceresana. Market study: carbon black, Ceresana, Constance, 2015.

[30] IARC/NORA. Identification of research needs to resolve the carcinogenicity of high-priority IARC carcinogens, International Agency for Research on Cancer, Lyon, FR, 2010.

[31] ICBA. What is carbon black? 2023. Retrieved from International Carbon Black Association: http://www.carbon-black.org/index.php/what-is-carbon-black.

[32] MCC. Application examples of carbon black, 2006. Retrieved from Mitsubishi Chemical: http://www.carbonblack.jp/en/cb/youto.html.

[33] NIOSH. Occupational safety and health guideline for carbon black potential human carcinogen. Atlanta, GA: United States Department of Health and Human Services: Centers for Disease Control and Prevention, National Institute for Occupational Safety and Health, DHHS (NIOSH), 1988. Retrieved from Center for Disease Control and Prevention.

[34] Vasudevan D, Gaddam RR, Trinchi A, Cole I. Core-shell quantum dots: properties and applications. Journal of Alloys and Compounds, 2015, 636, 395–404.

[35] Nanoscience and nanotechnologies: opportunities and uncertainties. The Royal Society & The Royal Academy of Engineering, London, UK, 2004.

[36] Samir TM, Mansour MM, Kazmierczak SC, Azzazy HM. Quantum dots: heralding a brighter future for clinical diagnostics. Nanomedicine, 2012, 7(11), 1755–1769.

[37] Mody VV, Siwale R, Singh A, Mody HR. Introduction to metallic nanoparticles. Journal of Pharmacy and Bioallied Sciences, 2010, 2(4), 282–289.

[38] Geraci C. Best practices for working with nanoparticles. In: 2nd Annual Massachusetts Nanotechnology Workshop Promoting the Safe Development of Nanotechnology in Massachusetts (pp. 5–8). Department of Environmental Protection, Boston, MA, 2009.

[39] Ellenbecker M, Tsai C. Interim best practices for working with nanoparticles. 2008, October. Retrieved from InterNano: http://eprints.internano.org/34/1/Best_Practices_for_Working_with_Nanoparticles_Version_1.pdf.

[40] OSHA. Working safely with nanomaterials. United States Department of Labor, Washington, DC, 2013.

[41] NIOSH. Progress towards safe nanotechnology in the workplace. Cincinnati, OH: United States Department of Health and Human Services: Centers for Disease Control and Prevention, National Institute for Occupational Safety and Health, DHHS (NIOSH), 2007.

[42] ISO. About ISO, 2023. Retrieved from International Organization for Standardization: http://www.iso.org/iso/home/about.htm.

[43] ISO. ISO/TC 229 Nanotechnologies, 2023. Retrieved from International Organization for Standardization: http://www.iso.org/iso/home/standards_development/list_of_iso_technical_committees/iso_technical_committee.htm?commid=381983.

[44] OECD. History, 2016. Retrieved from Organization for Economic Co-Operation and Development: http://www.oecd.org/about/history/.

[45] OECD. Nanosafety at the OECD: the first five years 2006–2010. Organisation for Economic Co-operation and Development, Paris, France, 2011.
[46] NIOSH. About NIOSH, 2023. Retrieved from Centers for Disease Control and Prevention: http://www.cdc.gov/niosh/about/.
[47] NIOSH. Strategic plan for NIOSH nanotechnology research and guidance. Cincinnati, OH: United States Department of Health and Human Services: Centers for Disease Control and Prevention, National Institute for Occupational Safety and Health, DHHS (NIOSH), 2010.
[48] SWA. Safety hazards of engineered nanomaterials. Safe Work Australia, Canberra, Australia, 2013.

Christie M. Sayes, James Y. Liu, and Matthew Gibb

5 Certification: Validating Workers' Competence in Nano-safety

5.1 Introduction

In today's workforce, it is essential to have tools for worker safety training [1]. Occupational Safety and Health Administration (OSHA) has traditionally taken on the responsibility of developing, directing, overseeing, managing, and ensuring the implementation of worker training in the United States. The training materials should help industry employees identify, reduce, and eliminate potential hazards related to nanotechnology, ultimately improving workplace health and safety [2, 3].

A successful nano-safety training tool requires a method to assess the trainee's content comprehension. Certification refers to confirming a person or organization's understanding of the specific technical content and serves as one way to accomplish this goal [4]. Some external review, education, assessment, or audit often confirms this. This chapter proposes what a worker safety training certification program might include, ensuring employees and employers collaborate to create a healthy work environment in the nanotechnology industry.

Training and certification in nanotechnology safety should include the following modules:

1. Introduction to nanotechnology: Defining terminology, describing benefits, highlighting challenges;
2. Occupational & environmental safety & health (OESH) life cycle paradigm: Designating occupational and environmental health issues specific to nanotechnology; and
3. Nanotechnology regulations and related standards: Explaining U.S. Environmental Protection Agency (EPA), Food and Drug Administration (FDA), Occupational Safety and Health Administration (OSHA), and Consumer Product Safety Commission (CPSC) regulations, regulatory requirements, and standards.

For further reading on module 1, see Chapter 1 of this book, as well as the following references: *Introduction to Nanotechnology* [5]. For further reading on module 2, see Chapter 3 of this book and the following reference: *Nanotechnology: Health and Environmental Risks* [6]. For information on module 3, see Chapter 6 of this book, as well as the following web pages: U.S. EPA (https://www3.epa.gov/), FDA (http://www.fda.gov/), OSHA (http://www.osha.com/), and CPSC (http://www.cpsc.gov/).

https://doi.org/10.1515/9783110781830-005

5.2 Definition of Nanotechnology for Training and Certification

The initial part of a nanotechnology worker safety certification course provides training on nanomaterials' structure, stability, and functional characteristics. The main aim is to give a broad understanding of nanotechnology and nanomaterials from the perspective of different stakeholders, including research, industry, manufacturing, trade, and regulatory environments. The subjects covered should include terminology and definitions, advantages and applications, difficulties, and potential hazards.

Several definitions of "nanomaterial" are used in domestic and international agencies [7–9]. The universal description states, *"materials of which at least one dimension is sized between 1 and 100 nm and for which has at least one unique physical or chemical property."* In the United States, the nanomaterial is *"any particle, substance, or material that has been engineered, purposefully produced and purposefully designed to be a nanoscale material and to have one or more dimensions in the nanoscale."* In Europe, the standard definition includes *"a natural, incidental, or manufactured material containing particles, in an unbound state or as an aggregate or as an agglomerate and where, for 50 % or more of the particles in the number size distribution, one or more external dimensions is in size range 1 to 100 nm".* In China, the working definition of nanomaterial is, *"a material which has a structure in the three-dimensional space in at least one dimension in the nanometer scale (1 to 100 nm) range of geometric dimensions or constituted by the nano-structure unit and a material with special properties."* And in Japan, the term nanomaterial *"refers to, among solid materials manufactured using elements, etc. As a raw material, a nano-object with at least one of the three dimensions of approximately 1 to 100 nm and a nano-structured material composed of nano-objects including matter composed of aggregated/agglomerated nano-objects."*

5.3 Occupational and Environmental Health and Safety Management

A nanotechnology certification course should offer training in risk management and occupational, environmental, health & safety. This training will provide an understanding of the best practices in handling nanotechnology and nanomaterials, focusing on risk and safety. The disciplines covered will include risk science (assessment, management, and communication), life cycle analyses, industrial hygiene, and medical surveillance.

According to the American Industrial Hygiene Association (AIHA) webpage (https://www.aiha.org/), *"Industrial hygiene (IH) is a science and art devoted to the anticipation, recognition, evaluation, prevention, and control of those environmental factors or stresses arising in or from the workplace which may cause sickness, impaired health, and wellbeing, or significant discomfort among workers or citizens of the community.*" Within the technology community, industrial hygiene professionals are responsible for examining the work environment and identifying any hazards and potential dangers, recommending safety improvement, leading research efforts to provide data on unsafe conditions in the workplace, proposing tools and techniques to anticipate and control the hazardous conditions, providing training and educational material to workers about job-related risks, collaborating with government officials in the development of safety regulations, and ensuring worker compliance as it relates to safety in the workplace [10]. Even though these tasks were not explicitly developed for nanotechnology, each can be applied to nano-safety in the work environment.

A risk management model, as it pertains to the nanotechnology workplace, can be defined as the assessment of engineered nanomaterials exposures and hazards of that exposure unique to an occupational worker or cohort of workers in conjunction with the recommendation and governance of procedures designed to minimize adverse impact [11]. In June 2007, the Environmental Defense Fund & The DuPont Company collaborated and produced the "Nano Risk Framework" [12]. The document *"describes a 'framework' for ensuring the responsible development of nanoscale materials. It establishes a process that companies and other organizations can widely use."*

The six iterative steps described within allow for new information to be incorporated even after implementing the steps within a company or institution. Step 1 describes the nanomaterial and its application in industry or commerce. Step 2 is data-rich and defines the process of developing the nanomaterial's physical and chemical properties, hazards, and associated exposures throughout the material's life. Step 3 involves evaluating risks based on information gained from steps 1 and 2; specifically, data such as identity and characteristics, the amount of material being generated, and probability of scientific, regulatory, business, health, and environmental risks are determined. Step 4 is the risk assessment piece, whereas step 5 is the risk management piece. Step 6 is prompt to review, revise, and adapt new information as it is developed in the company or institution's nanotechnology research and development efforts. Figure 5.1 depicts the steps of the framework and its iterative nature.

An occupational, environmental, health & safety (OEHS) model is a management process established within a company or institution concerned with the safety, health, and welfare of people engaged in the workplace or other work environment [13]. This model varies among industries, but the essential components include the following stages: anticipate, recognize, evaluate, control, and confirm. The nanomaterial is identified and categorized in the management process's anticipated step. After exposure, information about its biological effects is collected and banded based on a hazard statement. Hazard statements are assigned a unique numerical code that can be used as a handy reference when translating labels and Safety Data Sheets in other languages. Some documented characteristics include the sample's dustiness and potential exposure pathway based on its use.

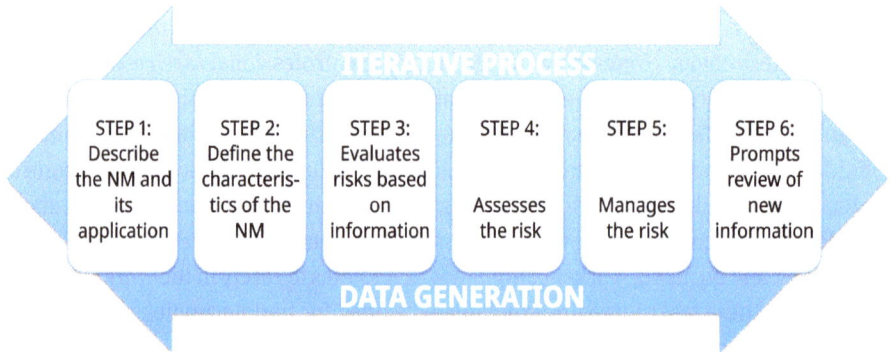

Figure 5.1: The nano-risk framework developed by the Environmental Defense Fund and The DuPont Company in 2007.

After the materials have been identified and categorized, the OEHS management process can recognize biological or chemical hazards associated with each sample. The recognition stage involves documenting the production, handling steps, and describing the work activities that might result in occupational and environmental exposure.

The next stage involves the evaluation of the nanomaterial health hazards. This is necessary to determine the best control procedures (i. e., implementing engineering control, personal protective equipment, or both). The evaluation includes an assessment of the effectiveness of these controls by validating the level of containment. Data gained from these continuous evaluations promotes an iterative process, where changes in control can be implemented without loss in the material production [14].

After evaluating the occupational health and safety management processes, the available data should be reviewed, and adjustments should be made to control occupational and environmental exposures. This management paradigm step matches the "level of control" with the "hazard banding schemes" employed. The level of control consists of selecting between the different types of personal protective equipment and engineering controls used by workers to ensure that exposures are minimized. Engineering controls are designed to reduce exposure to a hazard by substituting engineered machinery, equipment, tools, or process [15].

The last stage in the occupational, environmental, health & safety management model is confirmed. This confirmation process involves developing testing methods to measure real-time exposure and create a medical management system that reduces worker harm. This process is repeated each time a change in the work environment occurs. Any environmental variation also impacts potential exposures to new or existing hazards. Figure 5.2 depicts the five stages in the OEHS management cycle.

There are several reasons why workplaces need an OEHS management model. First, the program demonstrates employers' commitment to employees' health and safety; in other words, it outlines employer and employee accountability and responsibility. An effective program shows that workplace safety and business performance are compat-

Figure 5.2: The occupational, environmental, health and safety (OEHS) management model.

ible. Lastly, an implemented OEHS program sets safe work practices and procedures to follow to prevent injuries/illnesses.

Occupational hazards and risks in the workplace can be controlled by various methods, such as the precautionary principle, segregation of working staff members, mandatory yearly safety training, or quarterly equipment maintenance and process evaluation [16]. Arguably, the most effective risk management model requires a shared understanding of the nanomaterial of interest, followed by applying assessments and controls to achieve the appropriate balance between safety and productivity. Regardless of the method used, risks in the workplace are foreseeable and manageable.

5.4 Anticipating Hazards in Nanotechnology

Anticipating hazards in the nanotechnology workplace requires a thorough knowledge of the nanomaterial physicochemical characterization, its toxicological profile, information on hazard banding for the material or its chemical components, a human health risk assessment, and an ecological or environmental risk assessment [11]. Trainees should understand toxicology's role in determining the hazards associated with nanomaterials present in the workplace. They should be familiar with categorizing specific nanomaterials of interest according to their ability to cause adverse health effects and identify the recommended exposure limits for various classes of nanomaterials.

Because most occupational or industrial health professionals are experts in exposure assessment and containment, they are also trained to anticipate nanomaterial manufacturing operations' unique physical, chemical, and toxicological characteristics. Thus, a "cradle-to-grave" approach to developing a successful OEHS management model is possible. Trainees should be able to provide knowledge in the following areas: toxicology, industrial hygiene, and medical management. These knowledgeable individuals can streamline the compliance process by anticipating details of their OEHS needs well before assessment, thus minimizing potential hazards and risks.

The U.S. National Institute for Occupational Safety and Health (NIOSH) is a federal institute with two distinct responsibilities in preventing worker illnesses and injuries: conducting research and making recommendations. Specific to anticipating exposures in nanotechnology-related workplaces, aerosolizing particles on the nanoscale is the most significant concern to workers, employees, and regulators alike. To that end, NIOSH recommends a graded approach to assessing, measuring, and anticipating aerosol exposures [17]. Step 1 is to screen the area and process using particle counters and size analyzers. Step 2 is to collect samples at the source using filter-based sample collection for electron microscopy and elemental analysis. Step 3 is to collect samples from workers' clothing for electron microscopy and elemental analysis. Step 4 is to use less portable equipment and more sensitive aerosol sizing equipment for additional analytical studies and particle quantification.

Health banding provides a tool for industrial hygienists to assess exposure risk and risk management in the workplace [18]. This categorization is a logical system of classifying a nanomaterial of interest. Each material is placed into a health band based on its ability to cause harm. The data deriving from this assessment are acquired from a professional review of the current toxicological literature. By providing the relative hazard bands for the substances under review, the health band serves the working community in the qualitative aspects of risk management.

5.5 Recognizing Hazards in Nanotechnology

An occupational, environmental, health, and safety management program is meant to be carried out in all workplaces involved in the life cycle of a nano-enabled product, from its production and distribution to its use and disposal. The focus is on analyzing the specific nanomaterials of importance throughout this cycle (Figure 5.3).

Recognizing potential hazards in the nanotechnology workplace requires anticipating illnesses and injuries and asking the workers and employers behavioral questions. These questions should cover a wide range of elements, such as, but not limited to, fundamental questions of the OEHS program, hazard evaluation, exposure containment, communication, and education. The following list provides example questions appropriate for an OEHS manager or trained staff to recognize gaps in nanotechnology safety:

1. *Is there a demonstrated commitment to OEHS?*
2. *Do OEHS initiatives have both technical staff and senior management participation?*
3. *Does information exist relative to environmental fate & health effects?*
4. *Is appropriate technology implemented to minimize exposure?*
5. *Are facilities in place to contain and control exposures?*
6. *Are worker exposures continuously monitored and controlled?*

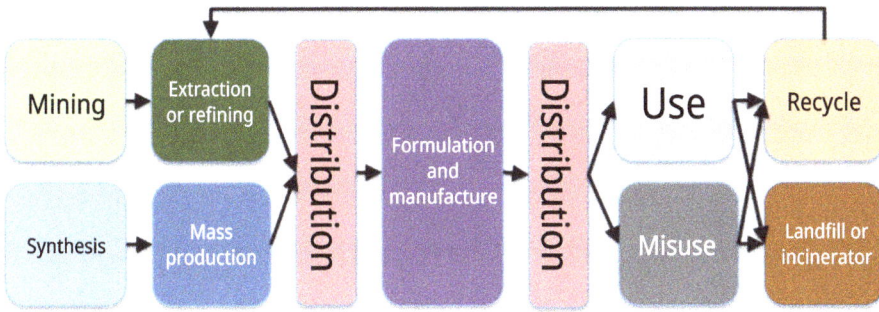

Figure 5.3: Recognizing potential hazards should be considered along the entire life cycle of the nano-enabled product.

7. *Are exposure and medical monitoring results communicated?*
8. *Is training at the appropriate levels available and provided?*
9. *Are SDSs and tech transfer information available?*
10. *Is the appropriate PPE selection available to all employees?*

5.6 Evaluating Hazards in Nanotechnology

To understand the hazards in nanotechnology workplaces, it's essential to anticipate and recognize them. However, it's equally crucial to evaluate whether a hazard event exists and the extent of the risk. Once health bands and the relevant nanomaterials are identified, the OEHS management model can help investigate potential exposures.

Evaluation includes determining an acceptable exposure level (AEL) for each nanomaterial. Once chosen, a "surrogate" is selected, and the work processes are simulated with this nonhazardous material. Adjustments are made based on the results of this effort, and nanomaterial handling can begin.

There are three unwritten laws of systems engineering used to evaluate the safety of a workplace [19]. First, there is an understanding that everything interacts with everything else, and the impacts of those interactions ripple throughout the system and should not be ignored. Second, every particle, vapor, fume, or liquid droplet deposits on a surface eventually. Third, never become so enamored with a new design decision that the consequences are not thoroughly thought through.

Containment is a large part of the evaluation stage and can be described as the action(s) of preventing a hazard from expanding into other work areas [20]. This takes on significant proportions in the workplace, particularly in nanotechnology laboratory environments, and should be considered in building design, process management, and research and development project work. Validation is the second significant component of the evaluation stage. It often refers to establishing documented evidence that a process or system can contain nanomaterials effectively and reproducibly when operated within set parameters. Used together, containment and validation provide the necessary information to evaluate an injury or illness in the workplace and prevent the injury or illness from happening again.

5.7 Controlling Hazards in Nanotechnology

Nanotechnology workplace assessment of activities should utilize the recognition and evaluation stages to generate recommendations for control. In other words, changes in infrastructure, institutional processes, or worker behavior can be made to control the conditions that exceed acceptable exposure levels. This level of control is generally achieved through a series of well-established industrial hygiene control solutions, such as ambient air dilution, ventilation, respiratory protection, isolation of nanomaterial, or other well-established technique. Because the OEHS management model is specifically designed for nanotechnology workplaces, control recommendations can be novel but are only implemented as necessary to achieve compliance requirements [21].

Controlling the hazards in a nanotechnology workplace requires identifying all nanomaterial handling processes. Once these processes have been identified, a control banding matrix can be laid-out and implemented (Figure 5.4). Control banding is used to manage workplace risks by matching a control measure (such as ventilation, engineering controls, containment, or personal protective equipment) to a range of "bands" of potential hazards (such as skin irritation, carcinogenic, or eye damage). Control banding is most effective in controlling exposure to the nanomaterial of interest [22, 23].

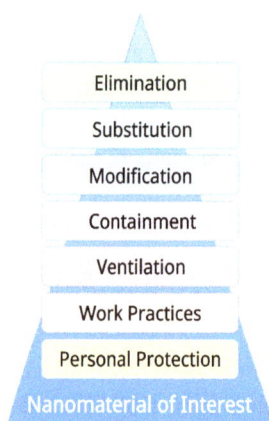

Figure 5.4: The hierarchy of controlling a nanomaterial exposure. Elimination is the first line of defense, and personal protection is the last.

Elimination and substitution.   These are the least practical control approach. It requires avoiding the use of nanotechnology in the workplace. Though there are reasons for eliminating the nanomaterial content from a product (such as the high probability of nanoparticles being released into the environment, the added benefits are marginal while the unknown risks are unacceptable, or health complaints from users of the product might be associated with the nanoscale component), this is not the preferred control measure in the nanotechnology industry. There is data to suggest that nanoparticles quickly aggregate, transform, or disintegrate once placed in environmental matrices, or the nanoscale component adds many benefits to the product, such as increased biodistribution in pharmaceutical applications and efficient catalytic conversation in petroleum applications [24–27].

Modification.   Modifying how a worker handles a nanomaterial of interest is a more acceptable way to control exposure. For instance, could a process be adopted to reduce airborne nanoparticles? Yes, nanoparticles can be provided and worked in a "wet state" to reduce the risks of inhalation exposure [28].

Containment and ventilation.   Gloveboxes are a containment system that keeps airborne nanoparticles inside a contained space during processing. Manufacturing facilities and research laboratories also use fume hoods, vacuum systems, clean rooms, laminar flow ventilation tables, and biosafety cabinets to control nanomaterial exposures [28]. However, each one of the containment solutions eventually needs to be opened to remove the product and waste or to clean or maintain the unit. Cautious work practices are necessary to minimize exposure during these specific activities.

Personal protection.   NIOSH recommends wearing hand, eye, and lung protection when working in the nanomaterials [21]. Limited data indicate skin penetration by these materials, but cuts in the skin may offer an easier path to internalization and the circulatory system. Disposable nitrile gloves are the most widely used, because they protect against various chemicals. Coveralls are designed to protect the skin. Using safety glasses for eye protection is also necessary. Goggles prevent splashes from getting into the eyes much better than safety glasses. Respirators prevent nanomaterial exposure to the lung [29]. The different types can be divided based on facial coverage: full-face or half-face. Full-face respirators have a significant design advantage: they go across the forehead. Building a respirator that goes across the forehead is much less problematic than trying to cover the bridge of the nose, where there is much more significant human variation. Not surprisingly, the greatest leakage occurs at the bridge of the nose.

5.8 Confirming Hazards in Nanotechnology

Managing occupational, environmental, health, and safety requires a unique process called "confirm." This involves reassessing conditions after a change in strategy or control upgrade. Nanomaterial manufacturing has unique characteristics, so slight variances can exceed established exposure limits. To prevent this, workplaces need an ongoing testing system to confirm any changes in exposure potential. This could include scheduling worksite assessments or implementing continuous detection processes [30]. The management model can stay current with the latest occupational health knowledge and practices by confirming.

Monitoring is a significant component of the confirmation stage. Nanomaterial monitoring can be classified as area, personal, or biological. Area monitoring measures nanomaterial concentration levels in ambient air before, during, or after a process or event is conducted [30]. Area monitoring can be used to establish background concentrations, trigger alarms in the event of elevated concentrations, and monitor long-term changes in air quality. Wipe sampling is another form of area monitoring, but it can also be used for personal monitoring. Instead of sampling a known volume of air, a known surface area (such as a countertop, instrument knob, patch of skin, or laboratory coat) is wiped. A new sample area should be wiped and used for each analysis to reduce the likelihood of cross-contamination. Biological monitoring measures the nanomaterial, metabolites, or biomolecules (such as enzymes of cytokines) in the blood, urine, or exhaled breath [31].

Biological monitoring can also be used as a metric for exposure. This type of monitoring confirms that the previous stages of anticipating, recognizing, evaluating, and controlling are not only being conducted but also reducing the uncertainty inherent in traditional exposure and risk assessments. Measuring the nanomaterial, metabolites, or biomolecules in blood, urine, or exhaled breath eliminates much uncertainty in estimating risk, because internal dose and response are directly available. It is a valuable tool for assessing human exposure to nanomaterials of interest, with measurements divided into exposure, effect, and susceptibility biomarkers. Lastly, biological monitoring provides unequivocal evidence of exposure when utilized as a part of an occupational exposure assessment [31, 32].

Employers conduct personal monitoring to assess the extent of their workers' exposure to hazardous materials at the workplace. This involves quantitative evaluation methods, such as personal monitoring and sampling, to determine potential hazards. The aim is to detect exposure levels that exceed the acceptable limits early on. The monitoring program also keeps a permanent record of your exposure. Personal monitoring devices, such as body badges and finger rings, frequently track exposure in high-risk areas.

5.9 Conclusions

This resource provides information on nanotechnology, including an occupational and environmental safety and health management model and a list of regulations and standards. Both technical staff and senior management need to be able to identify and control hazards in the workplace when working with nanomaterials. Workers should also be able to confirm proper containment and reduced exposure during all stages of the nanomaterial product life cycle.

5.10 Time to Reflect Questions

1. What are the three most critical components needed in a successful and effective nano-safety certification class?
2. List the five components to a traditional Occupational, Environmental, Health and Safety (OEHS) Management Method.
3. What term is used to describe the first line of defense against nanomaterial-related occupational exposure? What is used as the last line of defense against nanomaterial-related occupational exposure?

Bibliography

[1]   Glendon AI, Clarke S, McKenna E. Human safety and risk management, CRC Press, 2016.
[2]   Thomas K, Sayre P. Research strategies for safety evaluation of nanomaterials, Part I: evaluating the human health implications of exposure to nanoscale materials. Toxicological Sciences, 2005, 87(2), 316–321.
[3]   Roco MC. Environmentally responsible development of nanotechnology. Environmental Science & Technology, 2005, 39(5), 106A–112A.
[4]   Shulman L. Knowledge and teaching: Foundations of the new reform. Harvard Educational Review, 1987, 57(1), 1–23.
[5]   Poole CP Jr, Owens FJ. Introduction to nanotechnology, John Wiley & Sons, 2003.
[6]   Shatkin JA. Nanotechnology: health and environmental risks, CRC Press, 2012.
[7]   Kreyling WG, Semmler-Behnke M, Chaudhry Q. A complementary definition of nanomaterial. Nano Today, 2010, 5(3), 165–168.
[8]   Bleeker EA et al. Considerations on the EU definition of a nanomaterial: science to support policy making. Regulatory Toxicology and Pharmacology, 2013, 65(1), 119–125.
[9]   Maynard AD. Don't define nanomaterials. Nature, 2011, 475(7354), 31.
[10] Sellers CC. Hazards of the job: from industrial disease to environmental health science, Univ of North Carolina Press, 1997.
[11] Warheit DB et al. Development of a base set of toxicity tests using ultrafine TiO_2 particles as a component of nanoparticle risk management. Toxicology Letters, 2007, 171(3), 99–110.
[12] Defense E. DuPont (2007) Nano risk framework, Environmental Defense, New York, 2007. 2007.
[13] Robson LS et al. The effectiveness of occupational health and safety management system interventions: a systematic review. Safety Science, 2007, 45(3), 329–353.

[14] Jonsson P, Lesshammar M. Evaluation and improvement of manufacturing performance measurement systems-the role of OEE. International Journal of Operations & Production Management, 1999, 19(1), 55–78.

[15] Manuele FA. Prevention through design addressing occupational risks in the design and redesign processes. Professional Safety, 2008, 53(10).

[16] Danna K, Griffin RW. Health and well-being in the workplace: A review and synthesis of the literature. Journal of Management, 1999, 25(3), 357–384.

[17] Howard J. Director's Message.

[18] Paik SY, Zalk DM, Swuste P. Application of a pilot control banding tool for risk level assessment and control of nanoparticle exposures. Annals of Occupational Hygiene, 2008, 52(6), 419–428.

[19] Guldenmund FW. The use of questionnaires in safety culture research–an evaluation. Safety Science, 2007, 45(6), 723–743.

[20] Crowl DA, Louvar JF. Chemical process safety: fundamentals with applications, Pearson Education, 2001.

[21] Schulte P et al. Occupational safety and health criteria for responsible development of nanotechnology. Journal of Nanoparticle Research, 2014, 16(1), 1–17.

[22] Brouwer DH. Control banding approaches for nanomaterials. Annals of Occupational Hygiene, 2012, 56(5), 506–514.

[23] Johnson DR et al. Potential for occupational exposure to engineered carbon-based nanomaterials in environmental laboratory studies. Environmental Health Perspectives, 2010, 49–54.

[24] Banfield JF, Zhang H. Nanoparticles in the environment. Reviews in Mineralogy and Geochemistry, 2001, 44(1), 1–58.

[25] Reed RB et al. Solubility of nano-zinc oxide in environmentally and biologically important matrices. Environmental Toxicology and Chemistry, 2012, 31(1), 93–99.

[26] Banerjee T et al. Preparation, characterization and biodistribution of ultrafine chitosan nanoparticles. International Journal of Pharmaceutics, 2002, 243(1), 93–105.

[27] Kang J et al. Ruthenium nanoparticles supported on carbon nanotubes as efficient catalysts for selective conversion of synthesis gas to diesel fuel. Angewandte Chemie, 2009, 121(14), 2603–2606.

[28] Amoabediny G et al. Guidelines for safe handling, use and disposal of nanoparticles. Journal of Physics. Conference Series, IOP Publishing, 2009.

[29] Rengasamy S, Eimer BC, Shaffer RE. Comparison of nanoparticle filtration performance of NIOSH-approved and CE-marked particulate filtering facepiece respirators. Annals of Occupational Hygiene, 2009, 53(2), 117–128.

[30] Brouwer D et al. From workplace air measurement results toward estimates of exposure? Development of a strategy to assess exposure to manufactured nano-objects. Journal of Nanoparticle Research, 2009, 11(8), 1867–1881.

[31] Bergamaschi E et al. The role of biological monitoring in nano-safety. Nano Today, 2015, 10(3), 274–277.

[32] Simonet BM, Valcárcel M. Monitoring nanoparticles in the environment. Analytical and Bioanalytical Chemistry, 2009, 393(1), 17–21.

Walt Trybula and Deb Newberry

6 Understanding the Implications of Material Unknowns

6.1 Introduction

Upon reading the title to this chapter, the obvious question that occurs is "How can we understand what is unknown?" The answer is we cannot immediately or probably even within the time required to act to be able to make a considered decision. However, doing nothing is totally unacceptable and probably exposes the researcher, developer, manufacturer, and others involved to litigation. This chapter covers some situations where exact compliance is impossible, how to address risk avoidance, and how to prepare for the completely unexpected. As indicated in a previous chapter, metamaterials involving assembly/arrangement of nanoscale structures are considered to be included as nanomaterials.

It is *impossible* to address all possible situations involving nanomaterials. Why? There are a number of reasons, but only consider the volume of possible materials that need to be investigated. Based on work done under an OSHA Susan Hayward Grant [1], it is estimated that there are between 10^{200} and 10^{900} different materials and combination of materials that need to be investigated. To put that number in perspective, if the properties of one material per second had been determined since the creation of the universe, the total number of materials and combinations investigated would be less than 10^{18}! It is acknowledged that some of those materials and combinations may be immediately qualified as "undesirable," but the number of possible test candidates remain extensive, especially in light of the rapid development of 2-D structures.

Consequently, procedures must be developed for addressing situations where the material properties are unknown, which are not unlike firefighters moving into a fire of unknown cause, yet a general set of protocols exist. For example, in the case of safely dealing with an accidental spill of nanomaterials, there are certain general steps that can be taken dependent upon whether the nanomaterial is a liquid spill, a solid substance spill, or an airborne contamination. The efforts in Nanotechnology Safety (Nano-Safety) Education are addressing the training of students and workers to properly handle these situations [2].

This chapter builds on this previously mentioned work and focuses on developing an understanding of why these efforts are needed to ensure the application of safety in the development and application of nanotechnology. It not only is unacceptable to do nothing in addressing the needs of nanotechnology safety for people and the environment, it also opens organizations to potential litigation both now and in the future.

https://doi.org/10.1515/9783110781830-006

6.2 Background on Nanotechnology Safety Programs

As of the time of this publication update (April 2023), there have been only a few contracts awarded and completed that focus on Nanotechnology Safety and the related Education efforts. There is an ongoing METPHAST program addressing health and safety in nanotechnology and emerging technology. There are a number of Government training programs implemented within various federal departments. A certification in Nanotechnology Safety that is planned to be offered by Association of Technology, Management, and Applied Engineering (ATMAE). There are probably many individual efforts that have considered the issue of safety in handling nanomaterials, but very few that moved beyond the specific concerns related to a nanomaterial of interest. In many cases, the concern developed after there were many people involved in the research, development, and application of the particular nanomaterial. A representative example is Carbon NanoTubes (CNTs). CNTs were being used for strengthening materials, while providing lighter weight and found applications in Toyota bumpers replacing the metal that was then being employed. (The resultant was a stronger but lighter material that helped improve vehicle mileage.) Other applications included tennis racquets and baseball bats. Stronger, lighter products improved the performance of the users. This seemed to be a great new material. However, a question arose about the potential toxicity of the CNTs. The shape of a CNT (a narrow, straight tube) was noticeably similar to most forms of asbestos, which is considered highly detrimental to lungs if inhaled. Although the CNTs are considerably shorter, the issue was raised about the possible effects. A number of toxicity studies were done to evaluate the potential dangers but only after there were several applications and many participants involved in the use of carbon nanotubes

In the mid-2000s, there were some efforts to address the education of workers on the handling of nanomaterials. This was precipitated by the U.S. Occupational Safety and Health Administration's (OSHA) release of a guideline for the proper disposal of broken CFL light bulbs that contained micro amounts of mercury. However, there were larger amounts of unknown nanomaterials being employed in the absence of guidelines.

In 2010, a team from Rice University, led by Kristen Kulinowski with Texas State University as a subcontractor, was awarded an OSHA Susan Harwood Training Grant to develop and implement training modules on the safe handling of nanomaterials that targets the workers with the primary responsibility of safety in small-to-medium chemical companies.

The results were an 8-hour, seven-module training course that was validated at presentations during national meetings of experts. The modules were titled: 1) Introduction to Nanotechnology and Nanomaterials; 2) What Workers Need to Know about Nanomaterial Toxicology; 3) Assessing Exposure to Nanomaterials in the Workplace; 4) Controlling Exposure to Nanomaterials; 5) Risk Management Approaches for Nanomaterial Workplaces; 6) Regulations and Standards Relevant to Nanomaterial Workplaces; and,

7) Tools and Resources for Further Study. The intended audience for this course were people who already had a background in workplace safety but not aspects of nanomaterials.

In 2012, an NSF award was given to Texas State University, Jitendra Tate, Principal Investigator, and University of Texas at Tyler, Dominick Fazarro, Principal Investigator, to develop two Nanotechnology Safety Education courses. A development team was formed, which included Craig Hanks of Texas State, Robert McClean of Texas State, Satyajit Dutta of Texas State, and the first author of this chapter. A primary issue that faces education of students (or current employees) about the use of any emerging technology is to develop an appreciation for the potential impact of the technology—both benefits and risks. As Professor Hanks has stated in an abstract: "New knowledge, new techniques, materials, systems, and devices have brought new industries, new social forms, and new ethical challenges. One important aspect of technological societies is the intentional pursuit of change, of new technologies. This requires that responsible engineers have heightened awareness of the health and safety risks, ethical and social considerations and environmental and humanitarian implications of their work." [3]

Each of the two courses included nine modules and were designed to be offered either as a complete course or have modules extracted and inserted in to existing courses. The ability to add courses to curriculum based on current student credit loading is not always possible, but often module insertion is. The courses were developed, tested, enhanced, and retested. The complete courses were offered at UT Tyler and the modular selection employed at Texas State University.

The curriculum of the first course, *Introduction to Nanotechnology Safety*, consisted of the following modules: 1) What is Nanotechnology and Nanoethics; 2) Ethics of Science and Technology; 3) Societal Impacts; 4) Ethical Methods and Processes; 5) Nanomaterials and Manufacturing; 6) Environmental Sustainability: 7) Nanotechnology in Health and Medicine; 8) Military and National Security Implications; and, 9) Nanotechnology Issues in the Near Future. Referring to the primary issue mentioned above, it was determined that ethics, societal aspects, and environmental considerations needed to be introduced early in the course. When working on materials with unknown properties, there are many aspects of the development and production that need to be considered with thoughtful and diligent consideration of potential impacts, which may not be obvious until years after the initial applications.

The second course, *Principles of Risk Management for Nanoscale Materials*, contains the following modules: 1) Overview of Occupational Health & Safety; 2) Applications of Nanotechnology; 3) Assessing Nanotechnology Health Risks; 4) Sustainable Nanotechnology Development; 5) Environmental Risk Assessment; 6) Ethical and Legal Aspects of Nanotechnology; 7) Developing a Risk Management Program; 8) Results of case studies and research projects; and, 9) Hands-on experience in the Texas State Composites and Plastics Lab (designed specifically for handling nanomaterials). As with the first module, the inclusion of ethics and environmental considerations are a significant part of the educational effort.

The driving force behind this work is to provide an educated populace that neither fears nanotechnology nor assumes that there are no dangers. One can look back at the development of the asbestos situation. If there had been a knowledge of the dangers of asbestos to people, would the widespread application of the material have happened? Would it even have been considered? We don't want a reoccurrence due to lack of understanding.

The third program is developed by the University of Minnesota. With funding from the National Institute of Environmental Health Sciences Superfund Research Program, the Midwest Emerging Technologies Public Health and Safety Training (METPHAST) Program develops and disseminates web-based modules to educate and train a variety of learners about health and safety issues associated with emerging technologies. This is a multi-institutional collaboration among the University of Minnesota School of Public Health, the University of Iowa College of Public Health, and Dakota County Technical College.

The METPHAST Program's central objective for the current three-year funding period is to develop a comprehensive array of focused, web-based modules about nanotechnology health and safety, which can be used flexibly by instructors to create academic courses, continuing education initiatives, and individual lessons that serve the unique needs of different learners. This objective is being met by developing 20 one-hour, web-based modules and supplemental hands-on activities to train students and professionals to work safely with engineered nanomaterials.

The content can be presented by two academic courses. The first, *Introduction to Occupational Hygiene* is a 1-credit, semester-long course that uses 5 modules and accompanying activities to introduce students in science and engineering to occupational health and safety. This course is being piloted spring semester 2016. The second course, *Nanotechnology Health and Safety* will be a 3-credit, semester-long course that uses 15 modules and accompanying activities to provide in-depth training to occupational health and safety students and science and engineering students regarding nanotechnology health and safety.

The content is also intended for use by continuing education students and is arranged in a sequence of 6 units shown below.

– Awareness-level training for industrial hygiene professionals
– Operational-level training for industrial hygiene professionals
– Awareness-level training for other occupational health and safety specialists
– Operational-level training for other occupational health and safety specialists
– Awareness-level training for nanotechnology professionals
– Operational-level training for nanotechnology professionals

6.3 What Are Nanomaterial Unknowns?

Working with materials at the nanoscale for more than a quarter century, the nanoscale realm is still an embryonic technology, there are numerous unknown factors. Compounding this is the fact that many aspects of the nanomaterials involved must be considered. This list of characteristics is extensive; some of them are discussed within this section.

Current governmental efforts try to provide specific quantification for what constitutes a "nano" material. The majority of cases employ size as the determining factor. That size is 100 nm and smaller. Many things that are perceived as constants in the world actually change as the size of the material approached the atomic scale.

As has been the case for centuries, there are things that are known at the current time that were unknown only a short time before. For example, consider gold. Gold is a precious material that has been employed for ages. People working with gold are aware that its melting point is 1064.18 °C. While that is true for the bulk material, as the size of the gold nanoparticles decreases below 50 nm, the melting point starts to lower, and starts to drop significantly as the size becomes closer to 1 nm. Silver is another commonly employed material. As its size decreases to the 20 nm range, the silver nanomaterial is able to kill bacteria. The nanomaterial is employed in certain situations to prevent the development of bacterial colonies. Unlike antibiotics, bacteria "seems" not to build a resistance to the silver-based material. A problem arises, however because although silver nanomaterials can kill bacteria, it does not distinguish between good bacteria and bad bacteria, calling into question its use as an anti-bacterial material in some instances.

Carbon Nanotubes (CNTs), which are a nanoscale carbon material, exhibit strength while being lightweight. Usage of CNTs with polymers provides light weight replacements for automotive bumpers, which results in a stronger material than steel while being much lighter. The result is better vehicle mileage. Some may ask, however, about the lifetime of the bumper and what may happen as the car and associated materials are recycled or treated as waste. In other cases, one form of a single-walled CNT is conductive, whereas a slightly different orientation of the crystalline structure results in a semiconductor-type material resulting in different applications, which may have different pros and cons for material use. One of the characteristics of the nanoscale is exemplified by a slightly different form of carbon: graphene, which is an extremely strong sheet of carbon atoms. Graphene is often referred to as a two-dimensional material, because it can be only one atomic layer thick. Graphene is, simply stated, a carbon nanotube unrolled, but it has some totally different properties

The above examples show that the properties of materials at the nanoscale change from those at the bulk (macro) level. And the discovery continues; "new" properties of nanomaterials are continually being found. Therefore, the assessment of any system involving nanomaterials should be taken with the acknowledgement that we will be discovering more as the work progresses. The following is a list of some of the properties that need to be considered:

- Size: Aluminum nanoparticles become highly reactive in the range of 30 nm and are employed in certain types of rocket fuel.
- Shape: CNTs, which come in single wall and multi-walled configurations have different properties and when unrolled become graphene with additional different properties.
- Length: There is concern in parts of the medical profession that CNTs in excess of 5 μm long, and definitely over 20 μm, may cause similar reactions as asbestos.
- Thickness: The properties of graphene change as the number of atomic layers increase.
- Temperature: The melting point of gold and most metals starts decreasing as the nanomaterial size decreases below 50 nm.
- Purity: Graphene has significant electrical properties and is being evaluated for future electronic devices. Add hydrogen to the graphene and it becomes an insulator.
- Electrical: At dimensions below 50 nm, copper conductivity is dependent on the crystal orientation and grain boundaries of the nanomaterial.
- Color: Gold is known as a yellow material, but if it is put in suspension, the color of the suspension will change dependent on the size of the particles in solution. The red color in stained glass windows is dependent on a particular size of gold nanoparticle introduced into the glass during manufacturing. This characteristic was employed during the Middle Ages.
- Distribution: The nanomaterial may change behavior depending on the half-width of its distribution. Unfortunately, equipment to quickly and accurately measure large distributions is not currently available.
- Very small sizes: In the range of 1 nm, gold is a semi-conductor. Thirteen atoms of both silver and platinum have a magnetic moment. This is a new and unexpected property.
- Aging: There have been some indication that exposure to environmental conditions may cause some nanomaterial properties to be modified.
- Equipment Calibration: Equipment changes over time and when measuring properties of nanomaterials, even slight changes in equipment calibration can result in significant changes in the results.
- Equipment Design: Modifying (improving) portions of the measuring apparatus can result in inconsistencies of measurements.
- Atomic Level Shape: Cerium oxide less than 10 nm tends to take of the shape of a truncated octahedron with {100} and {111} faces; however, above 10 nm the structure shifts toward {111} octahedron [4].
- Multiple Material States: Materials in the nano realm are unique and different. Hochella [5] presents the fact that transition metal can exist in five *different states*, which are dependent on the item that the metal interacts with. The five states are: as hydrated atom; metal complexed in a small protein; metal adsorbed to surface on 1 nm mineral particle; metal adsorbed to surface of 20 nm particle; the same as previous except adsorbed to a 200 nm particle.

As can been seen from the above list, there are many possible influencing factors that can change the properties of the nanomaterials resulting in a complex assessment. There is insufficient data available to classify all nanomaterials and guarantee there will not be any hidden dangers.

6.4 Impact to the Public

In most cases, the general population receives information from the news sources. These sources provide descriptions of current and past events. Broadcast, print, and web media need to capture the audience's attention. How is this done? Usually it is some unusual event, whether it be a surprising success or tragedy. While there are also numerous other stories in the published media, this is not normally within the purview of the audience.

The assumption that must be made is that the first time the public will hear about and become interested in a subject, such as nanotechnology, is when there is some sort of a sensations event—either positive or negative—or even within the book and film industry. For example, nanotechnology-based discoveries that increase knowledge about diseases and aging often cause an increase in nanotechnology interest among the public. Similarly written material and movies, such as Spiderman and Spy Kids, can cause an uptick in interest in science in general and nanoscience specifically.

Of course, not all awareness is based on positive occurrences. Negative occurrences will typically involve some issue of nanomaterial toxicity.

There was one such occurrence that made the headlines in 2009. Reuters News Service reported on August 9, 2009 that seven women who worked (in China) became ill, and two later died after working with paint containing nanoparticles, while having insufficient protection from the material. (This was based on a 2008 study). The Chinese researchers who did the original investigation indicated the five surviving women suffered permanent lung damage. The claim was that this was the first case documenting health effects of nanotechnology in humans [6]. Clearly, this initial reporting presented the "risk" of nanomaterial use.

A further analysis was performed. A U.S. government expert said the study was more a demonstration of industrial hazards than any evidence that nanoparticles pose more of a risk than other chemicals. The women who were sickened on the job were spraying a paste containing nanoparticles in a very small, unventilated room, and wore gauze masks only occasionally. Toxicology findings from the lungs of the two women who died told the true story. The paste they were spraying accumulated and hardened in their lungs. There was a 30 nm nanoparticle in the center of the larger area of hardened paste, so the incident was described as a nanoparticle incident. More accurately, it was a failure of proper industrial protection for workers. But, it was originally reported as a nanomaterial incident. It is important to be able to provide facts that include not only materials involved but also aspects of human involvement and protection method-ologies.

This process requires public education. There are two separate areas that must be addressed. The first is the understanding of the health hazards of nanoparticles. Under-standing the hazards and degree of that hazard involve the material aspects covered in the previous section and also the method of nanoparticle introduction to the system. The following discussion assumes the human body as the "system." From what we under-stand today, nanoparticles can intentionally or unintentionally enter the body though 1) inhalation; 2) swallowing; and, 3) penetration through the skin. Obviously, nanomateri-als can be injected through medical procedures, but that is an intentional action.

– Inhalation can create a pulmonary inflammation. A persistent inflammation may cause a reaction that could lead to diseases as cancer or fibrosis.

– Swallowing or ingestion can permit the nanoparticles to transfer to other body or-gans by way of the gastro-intestinal tract. The effects are still being studied.

– Penetration through the skin (dermal exposure) may cause harmful effects locally within the skin or may be absorbed through the skin and circulated by the blood-stream. This can be caused by particles in the 1 µm (1000 nm) range, so nanoparticles would have no problem penetrating the skin. Again, the medical research requires specific particles and a long duration to determine the effects for most nanoparti-cles.

The second area which must be addressed and conveyed within the public arena is con-sideration of possible causes and issues with nanomaterials. These are areas of concern for the general public. Twelve of these are listed below.

– Toxicity of materials in many cases is known, and there are industrial safety precau-tions that are observed in industry. These precautions are adapted for the smaller sizes.

– Chemical dangers are also usually known, and steps are taken to prevent occur-rences; and if something occurs, steps are taken to control the situation.

– Fire is a danger in many ways. The containment of a fire is dependent upon the type of fire. This overall approach does not change based on size.

– Explosion can be a size-related issue. As mentioned previously, aluminum becomes extremely reactive as its size decreases. The need to contain the particles and pre-vent exposure to air (oxygen) is critical. This is a known issue and can be controlled by proper procedures.

– Nanoparticles in the form of free floating dust has concerns. Inhaling the substance can be an issue. The material could be toxic or could cause some unwanted internal reaction. This is one reason why breathing masks are strongly recommended for situations that may involve dust exposure.

– Electrostatic properties of materials from dust can enable the nanomaterial to cling to garments and be transferred outside the containment area for the materials.

- Size can be segregated into 3 categories. Nanomaterials less than 8 nm are normally considered minimal impact, since the current thinking is that these will pass the biology system in a person with minimal interactions. Particles in the range of 8 nm to 20 nm are a larger concern since the nanomaterial will have changed properties from the bulk properties and should have enough surface atoms exposed to be reactive. Over 20 nm in size, the nanomaterials may have impact; this issue is being addressed. Much of the needed information is not available.
- Shape is a concern. As mentioned earlier, the straight shape of CNTs has been likened to asbestos' shape. Since the needle-like shape of asbestos enhances its ability to penetrate into and accumulate in lungs, similar shapes have been considered dangerous. The one type of asbestos that is curved is not a capable of penetrating into lungs, but it is still considered part of the asbestos family and considered dangerous.
- Volume is a critical measure of the total amount involved and provides an estimate of the magnitude of the issue.
- Density of the material is important in that some of the heavier elements tend to concentrate in certain parts of the body and may continue to increase over time.
- Concentration is important. There are levels of exposure where different materials become a concern. Some nanomaterials can be toxic to certain organisms at levels that are a few parts per million or even lower.
- The separation of constituent elements due to natural degradation over time. One example is the separation of Cadmium from a Cadmium-Sulfide structure will provide a potential danger from the Cadmium in the environment.

The combination of all the factors listed above will determine if a hazard may be present but will also impact the degree of the hazard and its criticality.

The public also needs to be aware that nanoparticles are also created in nature. Forest fires and volcanic activity create nanoparticles. Even photochemical-based nucleation creates nanoparticles. Human activity creates nanoparticles. Welding can generate quantities of nanoparticles. Diesel exhaust, if untreated, will create nanoparticles. Even home gas cooking can create nanoparticles. Nanoparticles are the results of a variety of biological, chemical, and physical processes, some of which are commonplace.

The general public will have concerns if there are problems or issues with various materials. In some cases, there will be opposition to manufacturing that uses these materials in areas near populated neighborhoods. It is important to have adequate controls on all aspects of the application of nanotechnology and to ensure that workers are trained in the proper handling of these materials.

6.5 Risk Avoidance

There are always potential problems with any endeavor. The important objective is to proceed in a manner that will minimize the risks associated with a program/project.

Risk is a combination of the probability of a harmful occurrence versus the severity of possible harm. Any large project has a risk analysis associated with it. Building a large building or long bridge has the potential for a number of severely harmful or even deadly incidents. The goal is to minimize the incidents. The Golden Gate Bridge had a number of incidents that resulted in a total of 11 workers dying. This number was considered low compared to other projects at that time. The "low" number was due to what were extreme safety measures [7].

At the current time, there is insufficient data to create a classification of *all* nanomaterials that will guarantee there will be no hidden dangers or other surprises. The lack of information is that any risk assessment must proceed along the lines of identifying potential hazards and minimizing the risk. In many ways, this approach is similar to the training received by firefighting personnel. They are taught basic emergency analysis procedures and to approach each situation where the combustible material and other items are unknown.

The most efficient method of avoiding potential risk from nanomaterials is to control the exposure of the materials. The first is to contain the processes with controlled enclosures. This may range from a completely controlled facility to individual gloveboxes. Regardless of the method, there will be a need to have control of ventilation, both local and general. The usage of High Efficiency Particulate Air (HEPA) filters is recommended; these are mechanical filters employed to clean air before being returned to the workplace or exhausted. For particularly high exposure areas, air supplied respirators or filtering respirators that are properly fitted for the individuals must be employed.

Nanoparticles can cling to clothing. Consequently, protective equipment, such as clean room garments, are a wise investment. In situations where fully protective garments are required, there will normally be an airlock that has forced air streams to loosen an adhering nanoparticles, which are swept into a HEPA filtering system. Garments require a specialized cleaning process to remove potential residues. For situation with less potential for exposure, more traditional chemical lab garments have been used.

In all cases, there needs to be controls to reduce potential ingestion of nanomaterials. This starts with strict usage of gloves while handling nanoparticles. A face mask will prevent accidental hand to mouth touching. It is also important to have the workers understand the need for hand washing before eating and/or drinking.

With the concerns raised in the last few paragraphs, it becomes obvious that worker training is a necessity. There needs to be a series of training that is documented along with testing that indicates the understanding of the precautions that need to be taken. More facilities are capturing a baseline health of each worker as he/she starts employment in the nanomaterials industry.

Ideally, the organization has instruments that measure for potential contaminants and keep an ongoing record of the results. One company that has an excellent control system found an abnormal reading in airborne particle count. Fortunately, they also had on ongoing baseline outside their facility. The incident turned out to be caused by a very high pollen count in the area. There also should be a defined process regarding who has access to the nanomaterial secure storage areas and the frequency and times of access.

With precautions taken to minimize the risks to people and the environment in the manufacture and usage of nanomaterials, the need for the same principle regarding application of the products remains. This is a much more difficult situation since application of nanoparticles can take many forms. Developing a gold nanoparticle that is attached to a virus that is captured by a cancer cell, which permits killing the cancer has a limited possibility of large quantities of the nanoparticle getting into the environment. The more difficult application decision involves the use of quantum dots (semiconductor nanoparticles); the application of certain types of quantum dots reduce demand for generated electricity. [View screens, television, monitors, etc., use a significant amount of energy since they are normally on continuously. These same screens manufactured with quantum dots for the viewing area can reduce the power required by more than 50 %.] The reduced demand results in a reduction in pollution created by the source of generation. The potential problem is that if the quantum dots are released into the environment, they are initially harmless. However, over time they may separate into constituent elements, which could be toxic.

In spite of all the preparations and all the evaluations, there will be insufficient information to make a completely informed decision. Consequently, it is necessary to provide a means of making the best decision with imperfect knowledge. This is why there needs to be a focus on and understanding of ethics.

6.6 Ethics

Working with emerging technologies means working with uncertainty. Making choices that may take years to develop issues requires the inclusion of ethical choices. This is not an area that one normally connects with emerging technologies. Ethics impacts every industry and nanomaterial applications and use is not exempt. One needs to look no further than the recent Volkswagen scandal on "fudging" the emissions testing results to understand the need for ethics. Researchers at West Virginia University performed analysis of diesel emissions of vehicles from BMW and Volkswagen under a $50,000 grant from the International Council on Clean Transportation. The objective of the grant was to perform testing of the clean diesel cars outside a laboratory environment. Diesel emissions contain both carbon monoxide and NO_x, which have serious detrimental health effects. The approach employed by BMW and other manufacturers to reduce/eliminate the NO_x is to add a selective catalytic reduction system, which involves injecting urea into the exhaust. The resultant reaction turns the NO_x into nitrogen and oxygen. This is effective but requires adding a storage tank for the urea along with the associated parts required for the system to work. The Volkswagen development, was based on research from Toyota in the mid-1990s, which "periodically" used extra fuel to convert stored NO_x

into nitrogen and oxygen. The extra fuel reduced the fuel efficiency of the vehicle. By fine tuning the engine, they claimed they could meet both emissions requirements and still deliver mileage and performance [8].

The tests of the Volkswagen clean diesel cars found discrepancies that were 15 to 35 time worse than the Volkswagen published data [9]. (The other cars tested met the manufacturers' published results.) Eventually, this published data led to the California Air Resources Board investigation. That, in turn, led to both the CARB ad the EPA to raise questions regarding the large discrepancy directly with Volkswagen. The initial response by the company was that the differences were due to a technical issue rather than deliberate falsification. When the EPA threatened that it would not approve the 2016 clean diesel cars for sale in the United States, Volkswagen admitted they had inserted a device to provide low emission at the engine conditions employed to certify achieving environmental regulations. However, in order to provide the performance advertised, the emissions increased during typical highway driving. This is currently an ongoing problem for Volkswagen with numerous lawsuits by consumers and government regulators.

There are many more examples of problems caused by not having a background to consider the ethics of a situation. With nanotechnology and the development of manufactured nanomaterials, there are many more unknowns, which implies that a greater understanding and consideration is required for potential impact on individuals, society, and the environment. Consequently, ethics must be a part of any advanced usage of nanotechnology.

Technology development tends to increase in speed and not move toward an equilibrium. Technology creates new concepts, new applications, and opens even more opportunities for new actions and new goals. Progress is always moving on and creating changes and challenges as it develops. There are many examples of technology changes/challenges. Semiconductors started as a replacement for vacuum tubes switches. This evolved into integrated circuits, and then microprocessors (computers). The evolution of the cell phone to today's "smart" phone is another example.

As mentioned in Section 6.2, ethics education is an important focus of the overall education of nanotechnology development and applications. All emergent technology exists beyond current understandings and consensus. Understanding the ethical implications should provide an appreciation of the impact of the subsequent development of laws and regulations.

6.7 Government Pressure to Create Facts

When there is a concern raised by the voters, elected officials like to respond to their concerns. Politicians are not normally scientists or technologists. They react to the concerns being expressed. One example of this occurred in 2006. There was concern in the

City of Berkeley, California that research in nanotechnology could be dangerous and harmful to the citizens of the city. The City Council passed an ordinance that all materials that would be employed in nanotechnology research must have a Material Safety Data Sheet (MSDS) listing the characteristics of the material being employed, which includes demonstrated potentially harmful effects. At the time this was passed, carbon nanotubes (CNTs) were a particularly heightened focus of research.

The issue that arose is that there were no MSDS information on CNTs. The researchers needed the materials, but the information was non-existent. Research into the material properties was the reason for bringing the material into the area. This creates an interesting loop. You can't do research on the material, because the material properties are not available. You can't determine the material properties, because you can't bring the material into the research labs to test them. Go back and restart this loop. Obviously, no research would get done.

In many cases it is acceptable and permissible to employ closely related materials and their properties to quantify the probable characteristics of the material under question. In the case described, and per my understanding, the MSDS that was employed as a related material and therefore a substitute for the non-existent MSDS was the MSDS for graphite. Both are forms of carbon and exist at relative small dimensions. This satisfies the regulations requirement for a document where an exact document did not exist.

In this case, the City Council (government) required information that did not exist and would not permit experimentation to develop the data without having the data beforehand. Obviously, this is not possible. The government is trying to create facts that don't exist. The best you can hope for is a close approximation.

The characteristics of laws, regulation, and codes include 1) promoting minimal standards of conduct with the aim of ensuring safety; 2) providing lists of allowable and prohibited actions, with considerably more emphasis on the latter; and, 3) provide sanctioning and punishment for failure to meet the standards required [10].

The development of laws, regulations, and codes are always following the actual implementation of technology. Technology development is a high-risk opportunity. There are many more failures than survivors. And, the percentage of "winners" is a relatively small percentage of the total. Consider the life cycle of regulations and of companies. A typical time to get approval of a drug ranges from 7 to 10 years. This includes a phases of testing and evaluation. Some negative results are almost immediate. If one were to apply that time scale to testing the impact of a product going to market and compare that with the typical 18 month duration of a large number of startups, the product is developed, out in the market, and the company is out of business before their product was fully evaluated.

Since nanotechnology has been in development and application for a number of years, there are web sites that can be used to review nanotechnology-related regulations and guidelines. Some of them in the U.S. government are:

– National Nanotechnology Initiative – https://www.nano.gov

- Occupational Safety and Health Administration (OSHA) – https://www.osha.gov/dsg/nanotechnology/nanotechnology.html
- NIOSH – https://www.cdc.gov/niosh/topics/nanotech/default.html
- FDA – https://www.fda.gov/ScienceResearch/SpecialTopics/Nanotechnology/
- NIH – https://www.nih.gov/research-training/nanotechnology-nih

6.8 There Is No Place for Politics or Opinions

Scientific efforts need to be fact-based and not driven on the most popular approach that generates the most federally funded research dollars, or generates the most supporting opinions on mass media, or even a scientific theory that warns of impending crisis. When making decisions that have far reaching effects, it is imperative to have all the facts.

The ubiquitous Internet with various sources of possible information provides the opportunity to select information that may not always provide accurate information. Edward Muzio provides an interesting example of the possible quality of information [11]. His commentary discusses a mathematical problem presented on web post. The problem contained a series of mathematical operations on numbers that included parenthesis. With an engineering background, he enjoys solving problems, and then checking other answers. After he solved the problems, he was stunned on reviewing other people's answers. There was one person's statement "The answer is either 50 or 56. It depends on how you do the problem." Math is exact. There is no place for opinions. As he states: "To say that a mathematical equation is subject to a personal approach is to say that opinions are equivalent to facts. … It is simply not true. And being unable to differentiate your opinions from the factual reality in which they exist is, well, dangerous." This is true in math and is true in science. We need accurate data and resultant facts. There is a tendency to currently overlook that requirement.

In a newspaper commentary, Charles Krauthammer provided examples of accepted science that is later discarded or proven inaccurate [12]. One example is the federal government's 1980 warning of the bad effects of saturated fats. Public activists managed to persuade food companies to switch to trans fats. Now the Food and Drug Administration determined that trans fats are unsafe and ordered them removed from food. A second example is the medical "standard" that the average human temperature is 98.6 degrees F. In 1992, three researchers did an actual study and found that the average temperature is 98.2 F. In the span of time from the original effort to the latest, which was 114 years, what changed? Equipment capability? Methodology? Sample population? There is not a means to determine an answer.

In an excerpt from "State of Fear," Michael Crichton presents an interesting historical case [13]. "Imagine that there is a new scientific theory that warns of an impending crisis, and points to a way out. This theory quickly draws support from leading scientists, politicians, and celebrities around the world. Research is funded by distinguished philanthropies, and carried out at prestigious universities. The crisis is reported frequently in the media. The science is taught in college and high school classrooms."

This is not about global warming. He goes on to provide a list of the prominent people, who a century ago, supported the efforts. Major institutions performed research, and the efforts were supported by the National Academy of Science, the American Medical Association, and the National Research Council. "The research, legislation, and molding of public opinion surrounding the theory went on for almost half a century. Those opposed the theory were shouted down and called reactionary, blind to reality, or just plain ignorant."

This theory was eugenics, which postulated that a crisis in the gene pool was developing and was leading to a deterioration of the human race. The "superior" human beings were not breeding as fast as the inferior one, which included foreigners, immigrants, Jews, degenerates, the unfit, and the 'feeble minded'." The human disaster that built through the 1920s and 1930s culminated in the concentration camps of World War II. After that no one was or ever had been an Eugenicist.

What permitted the evolution of this "theory"? There was no rigorous definition of was meant by the terms. Worse is the fact that gene theory was barely understood. Without precise definition, people developed their own "definition." Furthering the cause was that there was no push back from the scientific community. The effort of this theory was an enabler for additional research funding.

In early 2016, there has been another method being employed to create scientific "fact." Senator Sheldon Whitehouse (Dem R. I.) 98th weekly US Senate speech calling for the U.S to employ the Racketeer Influenced and Corrupt Organizations Act (RICO) against organizations that are in disagreement with the climate change theory [14].

In September 2015, 20 climate scientists sent a letter to President Obama and Attorney General Loretta Lynch suggesting these official investigate using RICO laws against climate warming naysayers [15]. In effect, the climate scientists were asking the government to fine and/or jail other researchers who disagree with the climate scientists' positions. Unfortunately, the government threatened lawsuits, and hence created an environment hostile to independent investigations.

As Crichton concludes, he references Alston Chase's comment: *"when the search for truth is confused with political advocacy, the pursuit of knowledge is reduced to the quest for power."* What is needed is the development of knowledge that is disinterested and honest. That is the Scientific Method when followed in its intent.

The *Scientific Method* is a methodology for examining phenomena, developing concepts (hypotheses), testing them, analyzing the results, and proving or modifying the hypothesis as required. This process could start with a simple question of whether a nanomaterial were toxic or not. (That is developing a hypothesis. In this case, consider the potential impact of CNTs on lung tissue.) The next stage would be information gathering and analysis. (Asbestos is dangerous and 5 of its 6 types are needle shaped. CNTs are needle shaped.) With this information, one can develop an initial hypothesis predicting the outcome. (The needle shape of CNTs will be harmful to tissue found in human lungs.) Probably the most difficult step is to test the hypothesis through experimental procedures that others can duplicate and obtain similar results. (The first test might be something very simple to be able to observe/measure the reaction of tissue to CNTs.) The next stage is to analyze the resultant data. Based on the analysis, conclusions about the accuracy of the hypothesis can be made. At this point changes in the hypothesis can be made and new experiments run. Finally, a comprehensive report is developed and submitted for possible publication in a peer reviewed authoritative journal. Before being accepted, the submission will be reviewed by three to four experts in the field. Recommendations from these experts provide guidance on accuracy of the overall research, the thoroughness of the work, and the completeness of the submission. With the publication of the work, there is sufficient details provided that other researchers may duplicate the initial work and will obtain similar results.

Peer Review is an integral part of the process. The three or four experts review the submission for details of the research, the hypothesis, and the experimental means of testing the hypothesis. Completeness of the experimental details and the logic behind the conclusions from the findings are considered. The reviewers may recommend changes, modifications, additional information before accepting the work for publication. They can also reject the work due to insufficient completeness to the submitted documents. Historically, this has been the value of going through the work to have papers published. Unfortunately, there are pressures that have been compromising the peer review process [16].

With all the potential means of subverting the accepted scientific process, how is it possible to trust the published data? The most reliable sources are still the respected publications. With the Internet being a quick means of finding sources, the challenge is to ensure that the different accounts/reports of experimental results occasionally turn out to be multiple accounts of a single experimental report. Ideally, any referenced report that employs experimental parameters will also include data used and the equipment, if any, employed in the experiments. It has become necessary to more closely evaluate the information and observe any potential "strange" occurrences that are incorporated into the work.

6.9 Summary

From the material presented in this chapter, it should leave the impression that the effort required to obtain accurate information about new developments in nanotechnology is challenging. There are many possible dimensions to the application of nanotechnology, which we have only begun to explore. Without being able to have exact information, it becomes necessary to prepare for working with materials that are not really understood. Consequently, it is the researchers' obligation to learn as much about the material and its potential effects, while preparing for the possibility of surprising findings that require some type of preventative actions. In order to be prepared, a focus on maintaining and sustaining an environment that is friendly to both people and the environment in general is imperative.

Bibliography

[1] OSHA award announcement. http://news.rice.edu/2010/09/29/osha-bolsters-rice-based-safety-program-on-eve-of-buckyball-discovery-conference/.
[2] NSF Award Abstract. http://nsf.gov/awardsearch/showAward?AWD_ID=1242087.
[3] Hanks C, Fazarro D, Maleki S, Tate J, Trybula W. Encouraging Attention to the Humanitarian Dimensions of Emerging Technologies. IEEE 2015 Global Humanitarian Technology Conference Proceedings.
[4] Waychunas GA, Zhang H. Structure, chemistry, and properties of mineral nanoparticles. Elements, December 2008, 4, 381–387.
[5] Hochella, MF Jr. Nanogeoscience: From origins to cutting edge applications. Elements. December 2008, 4, 373–379.
[6] From Meridian Institute Nanotechnology Portal on Tuesday, May 18, 2010. http://www.merid.org/NDN/.
[7] Risk section. http://goldengatebridge.org/research/CheatingDeath.php.
[8] http://www.pbs.org/wgbh/nova/next/tech/volkswagen-diesel-emissions/.
[9] http://www.nature.com/news/the-science-behind-the-volkswagen-emissions-scandal-1.18426.
[10] Hanks C, Tate J, Fazarro D, McLean R, Trybula W. The Continuing Shock of the New: Some Thoughts on Why Law, Regulation, and Codes are Not Enough to Guide Emerging Technologies. 2015, NSF presentation.
[11] http://www.huffingtonpost.com/edward-muzio/opinions-arent-facts-math_b_7013256.html.
[12] Krauthammer C. Even science-fueled food fads prove to be wrong, pg A17, Commentary in January 2, 2016 Austin American Statesman.
[13] http://www.michaelcrichton.com/why-politicized-science-is-dangerous/.
[14] http://www.whitehouse.senate.gov/news/release/whitehouse-compares-climate-denial-to-civil-racketeering-perpetrated-by-tobacco-industry.
[15] http://www.sciencemag.org/news/2015/10/climate-scientist-requesting-federal-investigation-feels-heat-house-republicans.
[16] http://www.wsj.com/articles/hank-campbell-the-corruption-of-peer-review-is-harming-scientific-credibility-1405290747.

Evelyn H. Hirt and Walt Trybula

7 What is Considered Reliable Information?

7.1 Introduction

In this chapter we will address two complementary aspects of communicating information related to all things "nano." The first of these is the general concept of reliable information and its compatible concept of objective evidence. Second, we will address the overuse and sometimes conflicting use of the modifier and short-form or prefix "nano." How these two complementary aspects of communicating information are employed can complicate the usability of information, especially when considering the use of available research and impacts of nanotechnologies on worker safety and health. Please note that the reference to nanomaterials includes metamaterials that are developed through combinations of various nanomaterials.

The expectation of information reliability, sometimes referred to as data integrity, is always impacted by the user's (reader's) expected level of rigor and requirements to be applied to the information/data that is consistent with their plans for the purpose or intended end-use of the information. Possible negative impacts or losses attributable to unreliable information/data can be quite diverse but can include factors such as human injury; environmental damage; reduction in public safety; work-activity/project failure; financial loss, or reputation loss. As the severity or likelihood of these negative impacts or losses increases the expected level of rigor to be applied to the information/data would increase to manage/mitigate the assessed risk associated with its intended use within acceptable limits. In this context, the core attributes of reliable information are that the information is trustworthy, and the use of any information/results presented can be depended on to be readily available, consistent, and accurate (e. g., perform consistently under stated conditions).

The concept of objective evidence prevails across many disciplines (e. g., legal, business, scientific, engineering, quality, etc.). In all these disciplines the core attributes of objective evidence are that it refers to information (documented statement or record) based on facts, either quantitative or qualitative, which can be verified (proven) by an objective means (like based on observations, analytical tools, measurements, tests, or other forms of research). By its very nature objective evidence, even if it is completely unbiased, is information that is frequently open to discourse by those familiar with the field or topic (e. g., in peer reviewed journals), as it can be examined and evaluated by someone other than the one who presented the information.

Throughout history the concept of open discourse, or peer review, has been a critical concept in the distribution and sharing of scientific and technical information and a key factor in establishing sources of reliable information. Enterprises now operate in a global landscape that is highly influenced by the internet and big data. This impacts

https://doi.org/10.1515/9783110781830-007

the communication and usability of information. For example, the internet has been a boon to information/data publication and availability, but the quality and reliability of that information/data varies widely. Although many information sources found on the internet and elsewhere continue to utilize peer review and encourage open discourse, this practice can no longer be assumed by the user of information, unless some form of objective evidence is made available associated with the reliability of the information/data. Similarly, not all information/data on the internet is maintained or remains available over time. With an increase in online, scientific/technical publications, caution is required to ensure that references to research results are not in reality multiple references to a single published research publication. In other words, information/data found on the internet is best characterized as "consumers beware."

Use of the modifier and short-form or prefix "nano" usage has evolved over time, and changing to include its original usage as a measurement-modifier (i. e., denoting a factor of 10^{-9}) through its current common usage as an informal short form for nanotechnology and several iterations between. So, when reviewing information in support of obtaining knowledge about nano-safety or any other endeavor associated with nanotechnology, how certain can you be that from one source to another the usage of "nano" or nanotechnology related terms is comparable. One way to look at this is using the observation by David Owen [1], which he applied to the reproduction and communication of information: "Copying is the engine of civilization: culture is behavior duplicated. The oldest copier invented by people is language, by which an idea of yours becomes an idea of mine." So, consideration must be given to when do the use of "nano" or other nanotechnology terms constitute a comparable idea or information, whose language-usage may impact or affect the understanding of available research and its applicability. This is especially important when it comes to considerations of nanotechnologies impacts on worker safety and health.

7.2 Background on the Use of "nano"

As was previously mentioned the use of "nano" has evolved over time. In this section we will address some attributes to consider when interpreting the use of "nano" when determining information usage and reliability. Consistency of usage when nanotechnology and its short-form "nano" first emerged was more linked to the concept of small (e. g., a factor of 10^{-9} when compared to the original size of the item-type), then the attributes we associate with nanotechnology and its associated terminology today. Sometimes in these early days "nano" was employed more as a buzz-word to grab attention when writing grant proposals or pursuing contracts. In some cases, that usage had little or no relationship to the current standard usage of the terminology associated with nanotechnology as documented in International Electrotechnical Commission (IEC) and International Organization for Standardization (ISO) standards.

This was further complicated by liability issues that began to develop associated with "nano" materials, components, products, and processes. For example, at one time in the United States (U.S.) there was only one insurance company that would provide insurance for "nano" efforts. So once again, the significance of consistency of usage of "nano" began to drive a need for standard usage of the terminology associated with nanotechnology to support interpreting and litigating liability issues. Liability issues also drove the motivation to begin creating guidelines and regulations for the use and creation of "nano" materials, components, products, and processes. This also resulted in usage of "nano" often being eliminated and "emerging Tech" or something similar incorporated in everything from proposals to technical papers to items, and reducing the motivation to use "nano" as a buzz-word to just grab attention.

Prior to the emergence of nanotechnology standards, guidelines, and regulations, one of the most useful differentiations on the use of "nano" were that "nano" was not just a size reduction; it incorporated phenomena intrinsic to the nanoscale, which result in things like extraordinary mechanical properties and unique electrical properties, which were a function of the nanoscale. In other words, nanotechnology is an enabling technology that impacts electronic and computing, materials and manufacturing, health, medicine, energy, transportation, etc.

An example of phenomena intrinsic to the nanoscale is the memristor, one of the fundamental passive components along with the resistor, capacitor, and inductor. The memristor, a contraction of "memory resistor" was first theorized by Leon Chua in 1971 and was long considered an esoteric curiosity, because no one had managed to build one. However, it was in 2008 that advances in research yielded the physical reality of this missing non-linear two-terminal passive electrical component that was first produced by a team at Hewlett-Packard (HP) Labs and announced in *Nature* [2]. Presentations that year by R. Stanley Williams, a member of the HP team, highlighted that the enabling technology for this device is driven by scale. According to Williams [3]:

> It turns out that the influence of memristance obeys an inverse square law: memristance is a million times as important at the nanometer scale as it is at the micrometer scale, and it's essentially unobservable at the millimeter scale and larger. As we build smaller and smaller devices, memristance is becoming more noticeable and in some cases dominant. That's what accounts for all those strange results researchers have described. Memristance has been hidden in plain sight all along. But in spite of all the clues, our finding the memristor was completely serendipitous.

Currently there are standards being established and maintained, such as those by an ISO/TC229 and IEC/TC113 joint working group on Terminology and Nomenclature (JWG1). JWG1 is working on the creation and maintenance of standards to assist in the usage of the terminology associated with nanotechnology, such as those in the 80004 joint ISO/IEC Vocabulary Series. For example, as of March 2023, some under development and maintenance are:

– ISO/TS 80004-1:being revised *Nanotechnologies – Vocabulary – Part 1: Core terms*
– ISO/TS 80004-2:2020 Nanotechnologies – Vocabulary – Part 2: *Nano-objects*

- ISO/TS 80004-3:2020 *Nanotechnologies – Vocabulary – Part 3: Carbon nano-objects*
- ISO/TS 80004-4:2011 *Nanotechnologies – Vocabulary – Part 4: Nanostructured materials*
- ISO/TS 80004-5:5*Nanotechnologies – Vocabulary – Part 5: Nano/bio interface*
- ISO/TS 80004-6:2015 *Nanotechnologies – Vocabulary – Part 6: Nano-object characterization*
- ISO/TS 80004-7:2011 *Nanotechnologies – Vocabulary – Part 7: Diagnostics and therapeutics for healthcare*
- ISO/TS 80004-8:2020 *Nanotechnologies – Vocabulary – Part 8: Nanomanufacturing processes*
- ISO/TS 80004-9: 2017 *Nanotechnologies – Vocabulary – Part 9: Nano-enabled electrotechnical products & systems*
- ISO/TS 80004-10:yyyy *Nanotechnologies – Vocabulary– Part 10: Nano-enabled photonic components and systems* new peoject was abandoned 2019ISO/TS 80004-11:2017 *Nanotechnologies – Vocabulary– Part 11: Nanolayer, nanocoating, nanofilms and related terms*
- ISO/TS 80004-12:2017 *Nanotechnologies – Vocabulary– Part 12: Quantum phenomena*
- ISO/TS 80004-13:2017 *Nanotechnologies – Vocabulary– Part 13: Graphene and other two-dimensional materials*

Specifically, for example, ISO/TS 80004-9 includes definitions for "nano-enabled device," "nano-ink," "nanoscale device," "luminescent nanomaterial," etc.

These standards are supported by other ISO standards that support terminology and nomenclature development. Such standards will further the ability to communicate and evaluate information related to all things "nano."

So, what should you take away about the use of "nano" in terminology used for nanotechnologies and their applications?

- Ideally a source will include a definition of "nano" terminology, its uses, or it will site a technical standard, guideline or regulation defining the "nano" terminology they are using.
- Do not assume all sources of information you have access to use the term "nano" or nanotechnology-related terms in a comparable way. Check the context of the use of the term "nano" within a source, and then compare that context with the context used within other sources you plan to use.
- When publishing information be sure to establish the context of your usage of "nano" or nanotechnology related terms.

7.3 Information 'Fact and Fiction' the Dangers

Nano-safety, especially as it applies to products and worker safety and health, requires many things. These include knowledge of effects, understanding of particle behavior, toxic effects depending on the application, residual impact on the environment, etc. To achieve nano-safety, the ability to communicate reliable information/data supported by objective evidence is essential. This communication is the foundation to develop alliances to investigate issues and provide direction for the solution to any existing or potential problems that are uncovered. Communication is supported by addressing known nano-safety issues in a systematic manner to ensure that root causes are identified and investigated in an effort to develop effective methodologies to manage or mitigate existing situations and prevent foreseeable potential nano-safety issues. The following are some examples of implied dangers that have resulted when information "fact and fiction" impacted various situations resulting in various "nano" incidents or near misses.

7.3.1 Questionable Correlations in Chinese Workers' Deaths

One of the first widely reported nano incidents "causing" death involved Chinese workers. Reuters News Service reported on August 19, 2009 that seven women workers became ill and two later died from nanomaterials after working with epoxy containing nanoparticles. [4a, 4b, 4c] The facts turned out to very different, based on research work done in the United Kingdom (UK) by Ken Donaldson to find the real cause [5]. Chinese researchers reported seven young Chinese women suffered permanent lung damage, and two of them died after working for months without proper protection in a paint factory using nanoparticles.

The women who were sickened on the job were spraying a paste containing nanoparticles in a very small, unventilated room, and wore gauze masks only occasionally. In addition, the exhaust equipment had been broken and not repaired for over six months. Autopsies on the two women who died indicated that their lungs were coated with a material that had a 30-nanometer particle in its center. A U.S. Government expert said [6] the study was more a demonstration of industrial hazards than any evidence that nanoparticles pose more of a risk than other chemicals. UK toxicologist indicated [5] that the symptoms were more typical of chemical exposure. But the headlines had read that the women died from nanomaterials.

7.3.2 Questionable Identification of "Nano" Specific Dangers

A 2004 published DuPont report indicated that medical testing demonstrated that nano coal dust is harmful to human lungs [7]. This report only confirms what miners have known for years, which is that coal dust (any type) is extremely detrimental to one's health if it enters one's lungs. The report did indicate that there was some evidence that the nano variety might be more harmful. Since, coal dust is known to be deadly, publishing a report that highlights "nano" contributes a small amount of information, but permits sensational headlines, making "nano" in general to be perceived as dangerous.

7.3.3 Questionable Correlation of Carbon Nanotubes to Asbestos

A similarity in the shape of carbon nanotubes (CNTs) to asbestos creates another concern based on correlation reference. Asbestos is known to cause serious health problem. In fact, there are many advertisements on broadcast media that have lawyer advertisements for people or relatives of people with mesothelioma, which keep reminding people of asbestos dangers. So the similarity of physical shape between asbestos and carbon nanotubes, where both are needle-like, is a starting place to raise alarms about the dangers of CNTs. To further complicate the issue, the University of Florida published research in 2008 [8a, 8b] where an experiment was done to evaluate the effect of CNTs on a rat's intestines. These experimental results indicated that there were lesions observed and that had characteristic signs that were very similar to those seen in tests with asbestos (i. e., Mesothelioma). The general press picked up on these results and promoted the premise that CNTs will cause similar health issues to those caused by asbestos. Unfortunately, there are two issues that were not understood by these reporters. The first is that the CNTs were specially made to be 10 times longer than the typically manufactured CNTs. This extended length made the experimental CNTs roughly the same size as asbestos. The second fact is that the quantity (concentration) employed in the experiment was extremely high. Extreme doses do not represent typical situations and will cause abnormal experimental results, and are therefore not universally transferable to other applications or for comparison analyses. So, context matters and understanding context is one driving factor for citing references when publishing or discussing research findings.

If you need more clarity on the appreciation of the importance of "context matters," think of it this way: People have died due to consuming too much drinking (potable) water. Potable-water can be safely consumed in moderation. So is it proper to correlate the general drinking of potable-water by people as being an universal health hazard, just because consuming too much of it can lead to death?

7.3.4 Issues with NanoSilver Particles

Not all the information available is necessarily slanted in a particular direction. There have been a number of studies that have raised some very serious questions. It is known that nanosilver particles in the range of 20 to 30 nm have the ability to fight bacterial infections [9]. As a result, nanosilver is effective in this application and has been used in bandages. The issue is that the nanosilver does not decide that it will kill only bad bacteria, it will attack any bacteria. This becomes especially problematic when the nanosilver is not properly handled and gets into the environment; it can have very negative impact on some microorganisms. Some micro-organisms have susceptibility in the concentration of parts per billion, at which point it causes a 50 % fatality rate. However, some studies of chronic toxicity on fruit flies exposed to the 50 % toxicity levels have developed offspring with a resistance to the previously determined 50 % fatality level. [10]

This case demonstrates the validity of presenting and documenting all variables, potential risks, and impacts that are associated with a particular material application to establish the reliability of the information for potential users of the information. It also demonstrates a fair and balanced approach to discussing (documenting) key information in support of establishing objective evidence on a particular set of findings.

7.3.5 Over Generalization of Human Contact with Nanomaterials

Nano-titanium oxide has been in the headlines multiple times during the last seven years before 2016. There was a claim that the material (nano-titanium oxide) in sunscreen was dangerous and could cause significant harm [11a, 11b]. While it is an over generalization, the nano-titanium oxide is capable of causing significant damage. Due to its wide-spread usage, it can enter the human body through the mouth, skin, respiratory track, or other ways. The issue is that the material (nano-titanium oxide) can enter the blood stream and will settle in the liver, where it can cause inflammation. There is a significant amount of research available by various industries that support this conclusion and summarized by NIH. [12]. This currently available literature highlights that it is the accumulation of material that becomes the issue when it is used over time or in high concentrations.

The fact is that human ingestion/absorption of heavy metals tends to be retained by the body. This presents an interesting challenge. For example, some of the cancer treatments have employed gold nanoparticles attached to cells that are designed to be attracted to cancer cells. So this application yields two important items. First, the gold nanomaterial can be seen with medical imaging equipment and identify locations of cancerous cells. Second, that when the gold nanomaterial is attached to the cancer cells, an infrared light source can be employed to heat the gold to a very high temperature, which kills the cancer. These gold nanoparticles do accumulate in the body as do the nano-titanium oxide particles mentioned above.

The key question to consider is: What are the risks versus rewards for using a nanomaterial that is known to accumulate and be retained within the human body? Normally, it is possible to read through many articles published by reputable organizations. The long term sources of reliable information is tending to be enterprises such as government organizations, standards/regulatory bodies, or other organizations, whose funded long-term organizational mission includes supporting the generation and maintenance of nanotechnology related research, applications, and impacts. When organizational or activity (e. g., research) funding dries up (e. g., loss of government grants), the Internet sites, not affiliated with the type of organizations or bodies mentioned above, once relied on for information, usually become outdated quickly and possibly cease to exist.

7.3.6 Impacts of Litigation on Fact Finding and Misleading Correlations

Many times, there will be headlines in the news that such-and-such a company settled out of court for damages caused by their product by paying extremely large financial settlements. Paying these extremely large financial settlements does not prove guilt or even cause in these cases. Lawsuits require extra time of corporate management, including lawyers. It gets to a point that it is less expensive in both time and money to "settle" and get on with their real business rather than fight the lawsuit. This is especially true in a jury trial, where the "victim" appears to be a helpless, poor individual who was "victimized" by a ruthless company. The following example is from early 2016.

Johnson & Johnson been awarded a jury verdict requiring them to pay $73 million (U.S.) to the family of a woman who died of cancer, which was claimed to be caused by the material in the Johnson & Johnson talc-based Baby Powder [13a, 13b]. The jury found Johnson & Johnson liable for fraud, negligence, and conspiracy according to the family's lawyers. The case has created concerns regarding the safety of using talcum powder. Both Baby Powder and Shower to Shower products are made of talc, which are mineral rocks that contain magnesium, silicon, and oxygen, whereas some forms can contain asbestos [14a, 14b]. Everyone knows asbestos is a potential carcinogen. As pointed out in an article [15], both talc and asbestos are categorized as silicates, containing both silicon and oxygen. To further complicate the situation, the pictures of talc and talc with asbestos (long needle-like filaments) are available. So, talc with asbestos has the potential to be an issue. However, asbestos has been removed from any talc product since the 1970s. In addition, the victim was not sure which talc products she had used, and even if they were manufactured by Johnson & Johnson. There have been multiple litigations since the original version of this chapter, and the final verdict is not in.

The caution that the article on the talc verdict brings is twofold. First, the jury awards are not final, and even if it is settled without any contesting, there was no proof that the talc was responsible for the woman's death from cancer. There are a number of explanations in the references available that indicate there *might* be a link, but the probability is low. Second, the rapid reporting of an announcement like this will stay on the web regardless of the final outcome. This fact implies that as more and more articles, blogs, etc., are developed, it will be harder and harder to find out accurate information.

7.3.7 Sources with Conflicting Information

Even government sources may have opposing views. For example, this occurred in 2008 within the U.S. Environmental Protection Agency (EPA) when two offices within the EPA were providing different information.

"Manufacturers of nanoengineered products are getting frustrated by the uncertainties about the regulatory definitions of chemicals, materials, and products made with nanotechnologies. The U.S. Environmental Protection Agency's Office of Pesticide Programs (OPP) has come out with its definition of a "nanoscale material": "an ingredient that contains particles that have been intentionally produced to have at least one dimension that measures between approximately 1 and 100 nanometers," along with a new policy stating that an active or inert ingredient will be considered new if it is nanoscale. But the size-based focus of that definition is different from the one used by the EPA's Office of Pollution Prevention and Toxics (OPPT), which says size alone does not determine whether or not a chemical is new, and therefore subject to review under the Toxic Substances Control Act (TSCA)."

This example demonstrates the need to constantly review specifications and regulations. Initially, the EPA established a limit on the concentration of nanosilver. The limits were based on mature fish survival. As of 2016, the EPA has since changed the limits to reflect the survival of the more fragile embryos. In addition to the changing evaluation, results have been published [16] that demonstrate that the properties of the nanosilver changes over time. Also, the flies that were subjected to the testing and survived unharmed actually evolved to be immune to the effects of the silver nanoparticles [17]. Whereas this may seem unusual, one of the authors (Trybula) has seen similar effects with other nanomaterials [18].

7.3.8 Separating 'Fact and Fiction'

It is important to remember that there is no absolute source of totally reliable information. The vast number of possible nanoparticles (a low estimate is 10^{200} different particles) makes it impossible to provide an exact answer to every situation.

With the usage of "nano" or nanotechnology-related terms in applications covering a greater spectrum of uses, it is necessary to be able to evaluate the relevant and most current information from reliable sources. This challenge is addressed in the next section.

7.3.9 Source of Generated Information

The development of Artificial Intelligence as applied to professional documents provides an interesting challenge. Major Internet-based companies are developing "aids" that can gather information and generate text that seems reasonably accurate. One such "tool," ChatGPT, has been able to solve most of the "conservational" issues by the release of version 4.0. Whereas these tools may assist in writing, the original source of information might be unknown and in a number of cases are several years old.

7.4 Validity and Availability of Information Sources

One constraint on this discussion on the validity and availability of information sources is the assumption that books, in print or electronic, are readily identifiable and sourced via an internet search. At the time of the publication of this book, the tradition of sound peer and editorial review of books published by non-vanity publishers and their reputation in the publishing industry can still be used as an indicator of the reliability of the information they present. Therefore, this section will approach this topic from a broader perspective of information sources and how information is disseminated.

With enterprises now operating in a global landscape, what is the impact of using information sources that are highly influenced not only by their validity but by their availability via the internet? As discussed previously information/data found on the internet is best characterized as "consumers beware." It is up to the consumer of the information or data to apply critical thinking to objectively analyze and evaluate the source to form a judgement on the reliability of the information/data. On a positive note, more information is being made available over time by reputable institutions, professional societies, standards bodies, regulatory bodies, and governmental entities. However, there are some cases in which esteemed organizations/publications retracted previously published information. Therefore, it is important to double check sources for subsequent retractions. It is also advisable to evaluate all aspects of "scientific" reports that appear in highly political discussions. Selective choices of data can skew the conclusions with both sides using different portions of the data. In these cases, it is necessary to review the entire reports being referenced.

On a cautionary note, in addition to information reliability and integrity, not all information on the internet is appropriately maintained. The inherent cost to originate, review, and maintain information (including data) on the internet can result in sites that are shutdown or still exist, but the information is no longer maintained or updated. This is one of the reasons some internet sites and social media outlets supplement their income by accepting advertising or grant/agency funding to support continued operation. For example, International Council on Nanotechnology (ICON), a group concerned with risks and use of nanotechnology, was highly respected as a source of reliable information and collaboration but no longer exist. This most likely occurred when the grant funding that supported ICON was no longer available. Similarly, the Center for Biological and Environmental Nanotechnology (CBEN), formerly a department at Rice University, has a webpage that is still running to only provide resources to those who still need access to them; however none of the listed programs are still open for application. It is not predicable if CBEN will ever become active again or if the information on its website will ever be transferred to another host and maintenance of the information/data resumed.

For these reasons we have selected the following sample of resources and publications from enterprises (organizations) and entities that should have a viable financial model necessary to support the long-term creation, review, maintenance, and retention of reliable information/data for "nano" in support of research and considerations of impacts of nanotechnologies on worker safety and health.

7.4.1 Professional Societies Resources & Publications

– American Chemical Society
  – *Nano Letters*, http://pubs.acs.org/journals/nalefd/index.html
– American Society of Mechanical Engineers
  – Nanotechnology Institute, https://www.asme.org/ and search for nanotechnology
– American Vacuum Society
  – Nanometer Scale S&T Division, https://avs.org/about-avs/chapters/avs-divisions/nanoscale-science-and-technology/
– Institute of Electrical and Electronics Engineers, Inc. (IEEE)
  – IEEE *Xplore* digital library, http://ieeexplore.ieee.org/Xplore/guesthome.jsp, Provides access to millions of documents including research articles, standards, transactions, eBooks, and conference publications including but not limited to:
    IEEE Journal of Biomedical and Health Informatics (https://www.ieee.org/membership-catalog/productdetail/showProductDetailPage.html?product=PER171-ELE);
    IEEE Transactions on NanoBioscience (https://www.ieee.org/membership-catalog/productdetail/showProductDetailPage.html?product=PER191-ELE);
    IEEE Nanotechnology Magazine (https://www.ieee.org/membership-catalog/productdetail/showProductDetailPage.html?product=PER209-ELE);
    IEEE Transactions on Nanotechnology (https://www.ieee.org/membership-catalog/productdetail/showProductDetailPage.html?product=PER192-ELE);
    IEEE Pulse: A Magazine published by the IEEE Engineering in Medicine and Biology Society (https://www.ieee.org/membership-catalog/productdetail/showProductDetailPage.html?product=PER309-PRT);
    IEEE Journal of Translational Engineering in Health and Medicine (https://www.ieee.org/membership-catalog/productdetail/showProductDetailPage.html?product=ONL264)
  – IEEE Nanotechnology Council, https://ieeenano.org
  – https://trynano.org, Established as a resource for anyone interested in learning about Nanoscience and Nanotechnology.
– Materials Research Society
  – Nanotechnology Initiative, http://www.mrs.org/home/
– The Institute of Engineering and Technology (IET)

- *The Journal of Engineering*, http://digital-library.theiet.org/content/journals/joe/, Publishing articles covering a broad spectrum of engineering subjects, including micro and nanotechnology
- The Institute of Physics
  - *Nanotechnology*, https://iopscience.iop.org/journal/0957-4484

7.4.2 Government Sponsored Publications & Recourses

- United Kingdom (UK)
  - https://www.hse.gov.uk/nanotechnology/, UK organization for nano effects
  - https://www.safenano.org/, Established as a Centre of Excellence in Nanosafety in 2006 at the UK's Institute of Occupational Medicine (IOM)
- U.S. Department of Energy (DOE)
  - DOE N 456.1, *The Safe Handling of Unbound Engineered Nanoparticles*, (Canceled by DOE O 456.1) https://www.directives.doe.gov/directives/0456.1-CNotice/view
  - DOE O 456.1 Admin Chg 1, *The Safe Handling of Unbound Engineered Nanoparticles*, https://www.directives.doe.gov/directives-documents/400-series/0456.1-BOrder-admchg1
  - DOE P 456.1, *Secretarial Policy Statement on Nanoscale Safety*, https://www.directives.doe.gov/directives-documents/400-series/0456.1-APolicy
  - *Approach to Nanomaterial ES&H*, DOE Nanoscale Science Research Centers, http://science.energy.gov/bes/research/national-nanotechnology-initiative/nanomaterials-es-and-h/
- U.S. National Institute for Occupational Safety and Health (NIOSH)
  - *Approaches to Safety Nanotechnology*, NIOSH, http://www.cdc.gov/niosh/docs/2009-125/
- U.S. National Nanotechnology Initiative, http://www.nano.gov/
- U.S. North Carolina State University
  - https://www.rtnn.ncsu.edu/wp-content/uploads/sites/12/2019/05/Working-Safely-with-Engineered-Nanostructures_May-2019.pdf#:~:text=Working%20with%20nanomaterials%20at%20NC%20State%20University%20requires,and%20risks%20associated%20with%20using%20the%20specific%20nanomaterial. Nanomaterials Safety Guide

7.4.3 Other Information Resources

- *The Journal of Nanoparticle Research*, Springer, https://link.springer.com/journal/11051
- https://cosmeticseurope.eu/cosmetic-products/safe-design/, Cosmetics Europe – nanotechnology

- https://cordis.europa.eu/project/id/214478/reporting/it, Final report on addressing potential risks to human health and the environment
- http://nano-safety.org/, the Trybula Foundation's website addressing Safety in Nanotechnology
- http://www.nanosafetycluster.eu/, European effort to improve cooperation
- https://cordis.europa.eu/article/id/86196-new-nanotoxicity-database, Nano Health-Environment Commented Database
- https://www.qualitynano.eu, European funded effort to provide quality in nano-safety testing
- https://www.who.int/home/search?indexCatalogue=genericsearchindex1& searchQuery=nano&wordsMode=AnyWord, World Health Organization nano related publications

7.5 Summary Observations

The goal of the chapter was to present the general concept of reliable information and its compatible concept of objective evidence, and to address terminology employed with nano. Though there are reliable sources of information, the pressures of budget continue to reduce and change the number of established, active resources, and sources of information. The positive side is that there are many new sources maintained by enterprises (e.g., government sponsored efforts) and other entities. Unfortunately, nanotechnology is a dynamic field with constant changes. This results in many sites being slightly behind in information, while there are many site that purport to have scientific rigor that in fact are merely "wishful thinking" by the authors. The success of obtaining the latest details requires investigative searches for new developments. Some of the established sources of reliable information (including Internet sites) were provided in this chapter. These can be starting points, but further investigation is always beneficial and prudent to obtain the latest reliable information.

Bibliography

[1] Owen D. Making Copies. Smithsonian Magazine, August 2004.
[2] Strukov DB, Snider GS, Stewart DR, Williams SR. The missing memristor found. Nature, 2008, 453(7191), 80–83.
[3] Williams RS. How We Found the Missing Memristor. IEEE Spectrum. 2008. http://spectrum.ieee.org/semiconductors/processors/how-we-found-the-missing-memristor.
[4a] http://www.reuters.com/article/idUSN19481304.
[4b] Institute Nanotechnology Portal on Tuesday M. May 18, 2010. and their web site. http://www.merid.org/NDN/.
[4c] http://chem.pitt.edu/documents/201002111222210.Deaths%20Linked%20to%20Nanoparticles.htm.

[5]   http://www.nature.com/news/2009/090818/full/460937a.html.
[6]   http://www.reuters.com/article/us-china-nanoparticles-idUSTRE57I1Y720090819.
[7]   http://www.sciencedirect.com/science/article/pii/S1369702104000811.
[8a]   http://www.mrs.org/s08-abstract-mm//.
[8b]   http://www.scientificamerican.com/article/carbon-nanotube-danger/.
[9]   http://ehp.niehs.nih.gov/120-a386/ and https://arxiv.org/ftp/arxiv/papers/1101/1101.0348.pdf.
[10]   http://cen.acs.org/articles/89/web/2011/05/Fruit-Flies-Shake-Off-Silver.html6.6.
[11a]   http://copublications.greenfacts.org/en/titanium-dioxide-nanoparticles/.
[11b]   http://www.livestrong.com/article/283148-titanium-dioxide-sunscreen-safety/.
[12]   http://www.ncbi.nlm.nih.gov/pmc/articles/PMC3423755/.
[13a]   http://www.thedailybeast.com/articles/2016/02/24/can-baby-powder-really-cause-cancer.html.
[13b]   http://www.reuters.com/article/us-johnson-johnson-verdict-idUSKCN0VW20A.
[14a]   http://www.nanoshel.com/product/talc-nanoparticles/.
[14b]   http://usgsprobe.cr.usgs.gov/picts2.html.
[15]   http://www.cancer.org/cancer/cancercauses/othercarcinogens/athome/talcum-powder-and-cancer.
[16]   Lok CN, Ho CM, Chen R et al. Silver nanoparticles: partial oxidation and anitbacterial activities. JBIC
        Journal of Biological Inorganic Chemistry, 2007, 12(4), 527–534. https://link.springer.com/article/10.
        1007/s00775-007-0208-z.
[17]   Panacek A, Prucek R, Safarova D, Dittrich M, Richtrova J, Benickova K, Zboril R, Kvitek L. Acute and
        chronic toxicity effects of silver nanoparticles (NPs) on Drosophila melanogaster. Environmental
        Science & Technology, 2011, 45(11), 4974–4979. https://doi.org/10.1021/es104216b. Epub 2011 May 9.
        http://www.ncbi.nlm.nih.gov/pubmed/21553866.
[18]   Trybula personal communications from researchers at both Rice University and Boston University.

J. Craig Hanks and Emily Kay Hanks

8 Ethics and Communication: The Essence of Human Behavior

8.1 Introduction

The increasing evolution of nanotechnology and its resulting insertion into all aspects of everyday life and into life-making decisions (e. g., cancer remediation) provides challenges for the technology practitioner. Making potentially life-altering decisions without being able to know the potential impact on people and the environment necessitates some type of guidance. With products being developed and promoted over a short development period, even though full testing and evaluations may take decades, there is a need for some guidance. Understanding and applying ethics to situations enable some rational form of decision making.

A good engineer or technician[1] wants to do the best for clients and employers. Each also wants to do well for other stakeholders. What does it mean to be a good engineer, to be a good nanotechnologist? Clearly, one of the first and most important considerations is technical knowledge and skill. But, is that enough? Is it sufficient to have well-honed skills and have the most up-to-date information?

Although knowledge and skill are necessary, they alone are not enough to make a good engineer or technician, any more than knowledge and skill alone are sufficient to make a good physician or a good musician. A person might use knowledge and skill in the service of evil, or more likely fail to be sufficiently aware of the various responsibilities and virtues that are characteristic of engineering practice. Being a good engineer also requires meeting the standards and expectations of the profession, as set by governments and professional organizations. Professional codes of conduct, beginning with the Hippocratic Oath [1], establish rules and ideals to govern the conduct of members of the profession. Additionally, being a good engineer or technician requires effective communication with many parties, and failures to communicate effectively have important safety and ethical implications. Furthermore, safety, ethics, and communication concerning nanotechnology are made even more complex by the amount of uncertainty surrounding cutting edge technologies.

In engineering and technology, as in the other professions, neither knowledge and skill, nor staying within the bounds of law and regulation are sufficient to make one a good practitioner. As demonstrated during the Nuremburg Trials [2], by the Tuskegee Experiments [3], by the Kansas City Hyatt Regency walkway collapse [4], by the Challenger explosion [5], by Rachel Carson's *Silent Spring* and the ill-effects of DTD [6], and

---

1 For the purposes of this chapter, we will use the term "technologist" to refer to those who work with nanotechnology, including developers, researchers, technicians, and producers.

https://doi.org/10.1515/9783110781830-008

many other cases, engineering and technology also have ethical and communicative dimensions that go beyond following law and regulation and the skillful application of up-to-date knowledge. Even routine interactions and applications involve ethical considerations, because every stakeholder encounter involves the technologists' ethical duties, such as benefiting the client or employer and minimizing harm, and each encounter requires effective communication. Similarly, every technological decision engages the values of those involved, including clients, employers, regulators, and members of impacted communities. Being aware of these values and factoring them into decision making reflects ethical sensitivity and skill. It is only by also attending to the ethical and communicative dimensions of technology that one becomes a good technologist. In other words, a good technologist is a good communicator who is also ethically aware [7].

What does it mean to be ethically aware and to be an effective communicator? More specifically, what does it mean to be ethically aware about nanotechnology, and how does ethical awareness inform research, development, production, distribution, and use? Fortunately, we have over 2,500 years of thinking about ethics and communication that can inform these questions. In this chapter, we will (a) explain more about why knowledge and skill are not enough; (b) examine the relevance of major ethical traditions to technology; and (c) consider some of the communicative responsibilities of technologists working with emerging technologies. Each of these considerations sheds light on some important aspects of technical practice [8]. It is important to re member that dealing well with questions of ethics or communication is not merely a matter of applying any one theory to a situation, but rather "[w]hat matters is that the appropriated concept illuminates issues and provides conceptual resources useful in resolving practical difficulties." [9]

8.2 The Challenge of Ethics for Emerging Technologies

One of the central characteristics of contemporary technological society is the ceaseless and intentional search for innovation. This is a fundamental change from earlier, pre-industrial periods of human history, in which the human attitude toward technology was largely either preservative (to continue existing techniques) or to treat technology as something other, magical, or divine [10]. Contemporary residents of technologically advanced societies understand technology as a human product, and systematically seek to change existing technologies and create new ones. The development of new technologies aims to provide new material objects, new forms of efficient action, and thus allows new forms of social organization [11]. New technologies present not only new means for completing existing tasks, but create new possibilities, and thus new goals for human activity. This means that we have ever-new products, techniques, and goals, which consequently change individual lives, communities, nations, the international community, and nature itself. This also means that the presence of technology and constant change, intentionally sought, has come to be expected as the natural state of human existence, a taken-for-granted background condition. Additionally, new technologies have a power and a range of impacts—both spatially and temporally—greater than at any time in history, and also create situations of both great knowledge and great uncertainty [12].

Engineers, as architects of this new world, produce products and processes that impact the lives of all people in various, different, and often unexpected ways. Engineers and scientists [13] thus have responsibilities beyond developing and utilizing technical skills and knowledge. Consider, for example, nanomaterials. The past decade has seen a rapid growth in the development and use of nanomaterials in everyday objects (cosmetics and socks) and specialized items (wind turbine blades and geological sensors). Without some clear and intentional attention to ethical concerns, engineers (similar to other professionals) and the institutions within which they work, tend to focus on efficient performance and minimizing cost [14], or might avoid exploring nanomaterials out of caution. Realizing the full potential of this new technology demands guidance beyond the technical.

One reason to attend to ethical and social concerns is to mitigate criticism and resistance, such as happened in the case of genetic research, genetic engineering, genetically modified foods, and reproductive technologies in Europe and the USA during the 1970s to 1990s [15]. Another reason is that engineers and technologists have special responsibilities because of the role they play in developing and deploying new technologies [16]. A further reason is that accreditation requirements mandate that students develop "an understanding of professional and ethical responsibility." [17] Additionally, as we shall review below, emerging technologies are researched, developed, and often deployed before there is any social consensus about the wisdom of those technologies. A final reason, and perhaps the most important one, is that engineers and technologists are more than that; they are also citizens, parents, children, neighbors, and so on. Therefore, students and practitioners have many interests beyond the technical aspects of their work, many values beyond just avoiding harm [18].

8.3 What Does It Take to Be a Good Professional?

No one wants to be bad at a chosen profession. Who would seek work as a department manager, a director, or an elected official, with the aim to do the job poorly? No one. Rather, we desire to be good at what we do—to be a good manager, a good leader, a good director. We all want to be good professionals, but taken further, it seems reasonable to claim that we also want to be good parents, good siblings, good partners, good friends, and so on. Therefore, although our focus in this article is on professional lives, the ideas discussed are more broadly applicable. As noted at a recent workshop at the National

Academies of Engineering: ethics is foundational to being a good professional, to being a good engineer [19].

What does it take, then, to be a good professional? Part of the answer is obvious; it requires knowledge and skill. One of the distinguishing characteristics of professional roles is that they require specialized knowledge. Beyond specialized knowledge, professional roles necessitate the development of skill sets in order to translate knowledge into action. Often, these skill sets are also specialized. For instance, public managers might pursue professional education through a Certified Public Manager program or a Masters in Public Administration, whereas attorneys attend law school, and then continuing legal education. The National Society of Professional Engineers maintains a list of licensing requirements for PEs in each of the 50 states [20]. Most states require continuing education to maintain PE status, with many requiring regular ethics training. Some skills are generalized and necessary in most professions, such as skill in prioritizing tasks, but a surgeon or chef needs skills in cutting that a manager of a public agency typically does not. Engineers are particularly adept at using heuristic-based methods of problem solving [21]. A good professional is knowledgeable, and can deploy that knowledge skillfully and appropriately. This latter point is particularly important. To use knowledge and skill appropriately is to use them in the manner of an ethically mature and responsible person.

There are certain qualities or skills that characterize an ethically mature and responsible person [22, 23], a person who works not only to avoid problems but also to promote improved well-being [24]. The following four qualities should be encouraged in technologists and engineers:

- *Ethical sensitivity*: Awareness of the ethical dimensions of a situation, action, or institution.
- *Ethical judgement*: Capacity to evaluate the relevant ethical and factual decisions, to consult relevant sources of guidance, and to reach a decision about the best course of action.
- *Ethical motivation*: Desire to follow the decided course of action.
- *Ethical character*: Self-discipline to follow the decided course of action.

8.4 Technical and Procedural Knowledge and Skill Are Necessary, but not Enough

Although technical and procedural knowledge and skills are necessary to be a good engineer or technologist, they are not enough [25]. Not only might knowledge and skill be inadequate to the task (as when an issue arises due to a new technology), but knowledge and skill can also be used for bad ends.

Knowing how to do something does not tell us whether it should be done. If we always do what is right, then we need no rules and no laws. However, as we all know, humans sometimes act badly, and thus human societies and organizations benefit from the guidance of rules and laws. As noted by the great twentieth century Spanish writer José Ortega y Gasset, "Law is born from despair of human nature." [26] In a democratic social order, it is likely that bad actions are more often the result of a lack of knowledge, bad habits, and/or inattention than of outright maliciousness.

There are many possible reasons for bad actions, and many contexts in which they can occur. The following list outlines three situations:

A   Sometimes people use knowledge and skills for selfish reasons, for private gain for themselves or friends and family, and not for the good of clients, employers, or the community. Sometimes this prioritizing of self-interest can lead to bad actions.

B   And/or, people may use knowledge and skill for improper or bad ends because of a lack of guidance.

C   And/or, sometimes rules exist, but they are bad or unjust rules. They may be unjust in the context of a generally unjust regime (authoritarian regimes have laws and courts, and so on) or be unjust laws or regulations within a democratic society, as existed in the USA under the former Jim Crow laws.

Having rules is a typical response to "A" and "B." Rules can provide guidance and alert us to inappropriate selfishness in our own actions and in the actions of others. But, as we shall see in the next section, following the rules, even good ones, is not enough. As "C" points out, the rules may be wrong, and thus we need criteria for evaluating whether a rule is a good one or not.

8.5 Guidance from Rules Is Necessary, but Compliance Is not Enough

When faced with a question about what to do in a situation of uncertainty, individuals seek guidance, often in the form of rules. Rules take many forms. Some are formalized, such as the rules of chess. Others are formalized, but in practice have many variations, such as the rules of Monopoly. Yet others are formalized and have interpretation built into their application, such as the strike-zone in baseball. Many rules are informal, such as habits of applauding or not between movements of a classical music performance. Rules offer guidance about when and how to act, as well as when and how to make use of the knowledge and skills we have. The rules also tell us which knowledge and skills are relevant to the context in question. In this context, we are concerned with the rules that are relevant to work as a technologist. These rules are found in laws, regulations, and codes of ethics or conduct. Examples of the latter are professional codes, such as the *Code of Ethics* of the American Society for Public Administration, or institutional codes, such as the *Code of Conduct for EPA Staff* or the *Code of Conduct of the International Federation of Red Cross and Red Crescent Movement*. All of these (laws, regulations, codes) are types of rules. Codes of ethics sometimes have the standing of law (as in the case of state ethics codes for state employees). Additionally, laws are often justified by reference to the value or goods they promote or protect within a society [27].

Rules serve many purposes for the professional, [28] including the following:

– Informing clients, the public, other professionals, and practitioners of standards for behavior and decisions.
– Defining and promoting the image of a profession or institution, both internally and to the public.
– Providing support for practitioners.
– Serving as inspiration and guidance.
– Regulating behavior.
– Standardizing professional practice.
– Communicating expectations to professionals, clients, citizens, and government.

Law, regulation, and professional codes thus all provide some guidance to the safe and ethical development and use of new technologies. For technologies that are used in a variety of contexts by a variety of persons, "many of the ethical issues have already been identified by society." [29] In such a situation, we find increasingly complex and useful guidance to action codified in laws and regulations that represent an emergent social wisdom. This wisdom is arrived at through deliberation, trial and error, failures and successes, and through political, economic, and value debates [30]. Codes of ethics also are updated in response to technological changes. Professionals and society depend on the guidance of law, regulation, and codes to deal responsibly with existing technologies. The agreed-upon wisdom and guidance found in these rules can help practitioners avoid or resolve ethical problems.

However, rules are not the same as ethics, and following rules is not the same as acting ethically (even if it is generally an ethically good thing to follow just and good rules). Rules tend to provide guidance in reaching minimal standards and in avoiding some wrong. However, to be an ethical professional requires more, it also includes working to create a flourishing and vibrant situation, which is more than merely avoiding wrong [31]. For example, suppose someone works in an election office. The rules help that person avoid doing explicit wrong by unlawfully excluding anyone, and thereby decreasing the number of eligible and interested voters who participate. Ethics might also require that the person encourages more people to vote and promotes more active participation, because that is a good for a democracy. In this context, a fuller understanding of the ethical responsibilities of a professional points toward courses of action beyond avoiding violations of law. The following chart offers a brief comparison of rules, regulations, and codes with ethics.

A focus on rules, although necessary, can leave us with the idea that if we follow the rules we have done enough. This is sometimes referred to as an "ethic of technical compliance," a term coined by University of Miami Law Professor William Widen in his 2003 examination of the Enron case [32]. The ethic of technical compliance takes two

Laws, Regulations, and Codes	Ethics
Minimal standards	Aim at maximizing good, rather than minimal standards
Cover a limited range of previously encountered cases	Provides tools for evaluating new cases
Breaking a law or regulation can lead to criminal or financial penalties	Can evaluate whether laws or regulations are just
Breaking codes can result in loss of a license or job	Ethical failings are judged by individual conscience or by the community
Rest upon ethical principles and values, but do not evaluate those principles and values	Discussion, questioning, and evaluating ethical assumptions in order to obtain better understanding

forms. First, it can be articulated in the idea that if a person follows the rules, then that is enough. Of course, the rules may not be comprehensive, may be outdated, or may fail in subtlety or complexity. Even if they are generally good rules, they are likely to direct us only toward avoiding a wrong, not toward achieving a good. Remember the example above; we are technically compliant with the rules of work as a voting official if we do not unlawfully exclude anyone. But, if we believe this is all that is required (i. e., avoiding a wrong), then we do not do anything to increase democratic participation. The second sense of the ethic of technical compliance is when someone follows a narrow and technically correct reading of the rules, but does not act beyond that narrow understanding. This is commonly found within large organizations, and is a long-recognized characteristic of bureaucracies. Returning to the above example, if a voting official recognizes that more could, and perhaps should, be done to facilitate participation but does nothing to make that happen, because increased participation is not required by a narrow understanding of the job, or because the inertia of the organization is to continue what is already done and not rock the boat, then technical compliance occurs but the ethical good is missed.

Nanotechnology, like all new and emerging technologies, is creating possibilities and questions beyond established social consensus or ethical analysis. This places significant pressure on developers, researchers, and users to make thoughtful and morally responsible decisions. Faced with this sort of situation, where science and technology are operating at the limits of policy and ethics, an ethic of technical compliance is especially problematic. It is not possible to rely on existing guidance when none exists. Those who work with emerging technologies, such as nanotechnology, are among the most ethically important actors in contemporary society, holding both the possibility of great harm and the promise of barely imagined possibilities and benefits.

In his 1953 novel, *The Long Goodbye*, Raymond Chandler noted, "The law isn't justice. It's a very imperfect mechanism. If you press exactly the right buttons and are also lucky, justice may show up in the answer. A mechanism is all the law was ever intended to be."

Although it is necessary to have rules to provide guidance, and to help us guard against selfish motives, following the rules is not enough.

8.6 Considering Ethical Frameworks

After these preliminary considerations about the importance and limitations of having good rules and laws for guidance, how should an ethically responsible technologist act? How is an ethically responsible engineer to know what is the right, or better, course of action? Remember the four qualities and skills that we identified earlier as characterizing ethically responsible professionals. Two of these are *ethical sensitivity*, being able to identify an ethical issue or ethically problematic situation, and *ethical judgment*, being able to evaluate a situation or issue and reach a reasonable and defensible determination about an ethically appropriate course of action. Ethical frameworks help us identify the ethically relevant aspects of a situation (ethical sensitivity) and guide deliberation and evaluation (ethical judgement). Consideration of ethical frameworks can also help us think about how we can be morally good people, about our moral character, and thus point us toward a third essential characteristic—ethical character.

In the next three sections of this chapter, we examine the three most influential ethical frameworks in contemporary considerations of professional ethics. These frameworks originated over 2,000 years ago, in ancient Greece, and continue to in fluence law, policy, regulation, and individual behavior. These three frameworks help us consider different aspects of ethical decisions and actions; the character of the individuals and organizations involved (virtue ethics), the quality of the reasons given and the processes used, the value of persons (deontology), and the importance of balancing risk and benefit in the outcome (utilitarianism).

8.6.1 Deontology and Kant: Autonomy and Respect for Persons

The ethical approach of deontology focuses on fulfilling one's duties or obligations. The most important thinker in the deontological tradition is the German philosopher Immanuel Kant. Kant's moral philosophy is central to research ethics, medical ethics (both in clinical practice and in research), and professional ethics more generally. Deontology is an ethical approach concerned with acting rightly and for the right reasons, with recognizing and fulfilling one's moral duty, with acting according to good reasons, and with respecting persons. According to Kant, a good act is one that arises from a motive of duty and is rationally justifiable. When acting from duty, the agent is not concerned with personal desires, inclinations, or happiness, and therefore produces good actions that are neither influenced by external demands nor predicated on the outcomes they produce. Deontology focuses on the reasons or ethical motivations for action.

For Kant, the supreme principle of morality is what he calls the "categorical imperative," which he distinguishes from hypothetical imperatives. These latter imperatives tell us what we need to do in order to achieve a particular goal. For instance, if you want to lower your cholesterol levels, then you need to eat healthier foods. Technical codes can function as hypothetical imperatives. They tell us what to do in order to achieve a certain desired goal, for example, how to keep a boiler from exploding, or a bridge from failing. Unlike categorical imperatives, hypothetical ones are commanded as means to a particularly desired end. For Kant, moral imperatives must be independent of our desires or interests, therefore hypothetical commands might be instrumentally good, but they are not moral, because they are not done for their own sake. Much of what we do as members of large organizations, as members of work teams, as employees, or as consultants is similarly instrumentally good; it is good in a particular context with respect to particular goals.

According to Kant, what motivates one to act from duty is the internal logic of the proposed action or, in Kantian terms, the maxim. This is evident in the first formulation of Kant's categorical imperative, "Act only on that maxim through which you can at the same time will that it should become a universal law." Thus, good actions are those that a rational person could will as a universal law. In other words, the right action is one that follows an implicit rule that it would be rational to follow in every similar situation. Exceptions to the rule are then disallowed, because if everybody acted according to the exception, the rule would become inconsistent.

Kant's most famous example of a violation of the categorical imperative is that of telling a lie. A good example can be found in the history of medicine. Consider that empirical data from the 1950s to 1960s documented that clinical practice commonly involved lying to patients. An article published in 1953 showed that 69 % of physicians never told, or usually did not tell, their patients that they had cancer [33]; another study published in 1961 revealed that 90 % of physicians did not tell patients they had cancer [34]. This is in keeping with the fact that, for much of the twentieth century, the predominant principles in medical ethics were paternalism and beneficence. Because professionals possess specialized knowledge and skills, any professional might face the temptation to lie to a client, while believing it is in the client's interest. But, is this so? Is it ever justifiable to lie in a professional context? The deontological framework can help us think about this matter.

Lying takes many forms but, at its essence, it is any communication that knowingly intends to deceive another. This includes what ethicists call lying by omission, through not saying something. Suppose an engineer knows of a possible risk, such as the possible risks of inhaling carbon nanotubes. If the engineer is in a situation where he or she believes the risks are low and that others are not capable of rationally evaluating risk, it might be tempting to avoid talking about the uncertainty that accompanies new technology. This can be especially true when a professional believes there is a benefit that justifies the risk, and fears that the client might not take the risk. In these cases, a professional might want to bring the best outcome for the client, even if it means not identifying or explaining the risks. Not communicating about risks with those who might be impacted is a form of lying, and one that is not ethically supportable. As we will see below, such lying also fails to respect other people and violates their right to make choices about their own lives.

To lie to a client violates the categorical imperative, because a person of sound reason cannot wish this act to be a universal law. No reasonable person desires to lie always, or to be lied to always. This is true for Kant, not because of the problems it would create for the world if we all regularly lied; that would be a pragmatic influence on action. Rather, for Kant, lying is unacceptable, because willing it to be a universal law would mean that everyone would be lying all the time. This creates a conflict with the very idea of truth-telling and lying, and makes each impossible. Because the maxim of lying creates a logical conflict in its formulation, a rational person cannot accept it as a universal law.

Key to Kant's idea of ethical motivation is the role of reasoning. The physician who lies to a patient might justify the lie by saying that it was the only way to get the patient to do what he needed to for his health. This provides incentive for a lie, but it should not be mistaken for a reason [35]. Neither should it be mistaken for an ethical motive. In Kant's formulation, a good act has a reason that can be universally justified and that motivates the rational person to act from duty and against personal inclination and desire when these lead away from rational and ethical action. Consider the part of the Software Engineering Code of Ethics that specifies that software engineers will "not knowingly use software that is obtained or retained either illegally or unethically." [36] This provision states a clear rule that allows no exceptions, and it specifies a proper course of action, even if that action is not in the apparent best interest of an individual. Suppose you know that a coworker has access to a pirated piece of software or code that would help you complete a project, saving money and time. It might be tempting to use that code, but clearly not allowed by following the code of ethics.

Beyond the emphasis on identifying and adhering to rational rules, Kant's central contribution to professional ethics lies in his emphasis on individual autonomy and rights. His ethics is an ethics of respect for persons. People are rational agents and, as such, are never to be treated as mere means but always as ends in themselves. This is an idea well delineated in medical ethics, perhaps more so than in other areas of science and technology ethics. Respect for persons means that professional paternalism—making decisions for another—is unethical when faced with a person capable of making decisions about her or his life or goals. This idea is firmly established in clinical and research bioethics, in the prioritization of autonomy, and in practices of obtaining informed consent. It should also be central to all professional ethics, including the practice of engineering [37]. Obtaining informed consent is a way to ensure that people are treated as ends and that the proposed action (i. e., treatment, diagnostic procedure, enrolling as a research subject, how a new bridge will impact a community, whether a new nanomaterial should be used for cookware or clothing, and so on) is compatible with an end the patient or client has set. This is the case even when the goals and values of the patient or client are not those of the professional(s), or even of the majority of patients or clients [38].

This was not always the case, as the above-cited evidence concerning lying to patients shows. Beginning with the Nuremburg Code [39] and continuing through the World Medical Association Declaration of Helsinki [40], respect for patient autonomy has been established as one of the key principles in bioethics [41], and clinical practice reflects this change in patients' status. One marker of this change is data concerning whether or not patients are told that they have cancer. Unlike the customary practice two decades earlier, a study published in 1979 showed that 97 % of physicians reported routine disclosure of such diagnoses [42]. This finding reflects advances in several areas, such as the ability of clinicians to provide better care to patients with cancer. It also represents an increased recognition of the need for doctors to respect the values of their patients by providing them with the necessary information to make good decisions, as well as the increased dialogue between doctors and philosophers that began in 1970 [43]. It is also a step away from the unjustified paternalism of the past, which continues to characterize the practices of many professions and professionals.

Of course, respecting the autonomy of clients should not be understood as simply giving information, and then letting the clients fend for themselves. On the contrary, it requires that professionals foster autonomous decision making by disclosing information about clients' conditions and goals in appropriate ways, about reasonable alternatives for managing or achieving them, and about the benefits and risks of these alternatives. Further, clients should be free to select among these alternatives, and they should not be substantially controlled in either their process of decision making or in their final decision [44]. Moreover, professionals also enhance autonomous decision making by being aware of, and respecting, the values and desires of their individual clients, as well as by helping clients to overcome their sense of dependence.

8.6.2 The Pursuit of Happiness: Utilitarian Ethics

Attending to the goals that a client has set considers individual preference and desire, and in some ways is consistent with some aspects of utilitarian theories of ethics that demand that the agent act in ways that maximize the good. Contrary to deontological theories, however, utilitarianism requires that in evaluating the rightness or wrongness of an action or practice we look at its consequences rather than at the ethical motives or ethical character of the agent.

The most influential formulation of utilitarianism is found in the works of nine teenth century British philosophers Jeremy Bentham and John Stuart Mill. Although their theories are not exactly alike, both agreed that the purpose of morality was to promote human welfare. The basic principle of utilitarianism is what Mill called the "principle of utility." According to this principle, actions are right when they maximize pleasure and minimize pain [45]. The individual agent who is weighing two possible actions to decide which course to pursue must ask which act will produce the most happiness or pleasure. For example, weighing the benefits and risks is one way that an individual might go about deciding whether to consent to a medical procedure.

Utilitarian theorists understand the desirable outcome as the situation that produces the maximal balance of happiness over unhappiness. This means balancing costs and benefits for all affected, and seeking to increase pleasure or happiness and minimize suffering for the greatest number. Contemporary Utilitarians call our attention to the often neglected portion of the process—relieving or avoiding suffering, especially unnecessary suffering [46]. They argue that one of the most important of human interests is avoiding pain.

Importantly, utilitarianism is not a defense of crude self-interest. The point is not to maximize one's own happiness at the expense of other people's happiness, but to give equal and impartial consideration to the interests of all affected parties. Thus, when making decisions about how to proceed, one must not allow one's own happiness to weigh more heavily than the happiness of others. Similarly, when evaluating the consequences of a particular course of action, both indirect and long-term consequences, and not just direct and short-term outcomes, must be taken into account. Utilitarianism is, thus, a form of cost–benefit analysis, and requires a type of reasoning familiar to most people in technologically advanced societies.

Utilitarian ethics is often employed in considerations of the allocation of resources, such as in determinations of healthcare policy. Cost–benefit analyses of resource allocation are an example. This follows the goal of the original utilitarian thinkers—to develop a social ethic. Clinicians take up the utilitarian focus on outcome and apply it to the individual patient and family. In this way, clinical ethics sets a goal of maximizing happiness for individuals, who must define happiness in their own terms and are expected to consent to or decline interventions based on that definition.

Utilitarianism is also used every day in engineering and technology. Utilitarian ism is present every time a cost–benefit analysis is performed. Utilitarianism is a powerful and important ethical perspective. However, the goal of maximizing happiness presents some problems.

One problem with attempts to maximize happiness is that when confronted with a new situation, or one that is especially puzzling, people can be at a loss. Utilitarianism recognizes this, and proposes that the way out is to consult those persons with appropriate knowledge and experience to serve as guides to ethical judgement. Thus, in medicine, ethics consultations can help patients and physicians understand and evaluate the various choices they face. Unlike the Kantian approach, which aims for a universally correct answer, this approach aims to provide the best answer we can have at the time of the decision.

In a professional setting, the combination of Kantian ethics, with its focus on the autonomous individual, and utilitarian ethics, in which the right action is the one that maximizes happiness and minimizes suffering, illustrates how appropriating concepts from more than one theoretical approach can help address clinical situations. Professionals should respect clients' decisions, and those decisions are often guided by subjective evaluations of happiness and suffering. We might also consider the convergence of the two approaches around the matter of confidentiality of personal information. The deontological approach would value clients having control over their life and goals, including information about their lives, and thus respect for persons requires confidentiality. The utilitarian approach would argue that clients are more likely to seek advice and services, thus maximizing a good, if they can expect confidentiality [47]. Thus, both views argue that we should respect the privacy and confidentiality of clients, but point out different reasons for doing so.

8.6.3 Virtue: Character and Practice

Virtue ethics is sometimes characterized as more descriptive of professional practice and as having more in common with the practice of an ethical professional provider than deontology or utilitarianism [48]. Virtues are aspects of the character of a person put into practice. For instance, loyalty, fidelity, compassion, and benevolence are some of the virtues associated with health care practitioners. Importantly, however, virtues are both these inner states and modes of acting and can only be cultivated and realized in practice through nurturing virtuous relations with other persons. We can learn and develop virtues as we learn to play a musical instrument, or play a sport, or cook, or fly a plane; we need understanding and practice. Thus, the virtues of good technologists are not only their inner values, but also their relations with clients, employers, families, communities, and other practitioners.

Contrary to deontological and utilitarian theories that focus on evaluating moral actions, virtue ethics focuses on the ethical character of the agents who perform the actions. The point is not simply to determine what action to perform, but what kind of person to be, how to become a virtuous person.

Some virtues are necessary for moral action in general; others are special responsibilities for some, such as those in a profession [49]. For example, everyone with enough understanding of the world to appreciate its dangers can recognize that some degree of courage is required to do almost anything. Courage is needed by those who travel (to fly in an airplane or ride on a bus), by patients (to submit to medical tests or surgery), by physicians (to treat patients with communicable diseases), and by engineers who develop and produce new ways of doing things. We certainly want professionals, like everyone else, to be courageous people. Yet, without some special justification, there is no reason to count courage as a special feature of professionalism. Yet, the duty to consider the public good can demand that courage require more of technologists than of others. Alternatively, although sympathy is a virtue for all, it is especially applicable to physicians, whereas the virtue of prudence (thoughtfulness and avoidance of unnecessary risks) might be especially important for engineers [50].

Virtue ethics began with Aristotle [51]. One of the best contemporary accounts of how the virtues can work in modern society is that of Alasdair MacIntyre, who describes the virtues as socially embedded and as developed in relation to a practice that has a social history and goods internal to it [52]. He defines a practice as "... any coherent and complex form of socially established cooperative human activity through which goods internal to that form of activity are realized." [53] In the performance of these functions, activities, and attitudes one tries to achieve those standards of excellence that are appropriate to that form of activity.

When we apply MacIntyre's concept of practice to engineering, we can define engineering as the total set of skills and attitudes that are applied in the context of a particular goal or project, with the intention of providing good services or products (the goal) to another (person, community, organization). Noteworthy in this definition is the goal-oriented character of engineering practice. Whatever engineers do, for the ethically responsible engineer, it must always be related to the final goal. The goal of technology is the improvement of human life, and thus implies care for persons and communities. Virtue theory understands good care in the broad meaning of the word, that is, focused on the physical as well as the psychological, relational, social, and moral.

Importantly relevant to the practice of engineering are the notion of internal goods and the focus on the social and historical nature of practice. Goods are internal to a practice in two senses: (1) the goods can only be described in terms of the practice, and (2) they can only be acquired and recognized by engaging in the practice and having the associated experiences.

Unlike Kantian duty, internal goods cannot be abstracted from the particular experience and made into external rules. In the example of lying to a client, the Kantian analysis requires that professionals disengage from their experience and form a rule abstracted from the particular experience that can be applied in all situations that involve lying. In contrast to this, a description of the technologist's experience in a similar situation with a particular client exposes the values that are at stake.

Moreover, in virtue ethics, acts are best understood not in relation to abstract implicit principles of action, but as a socially embedded activity that takes on a narrative form with a past, present, and future. In committing to working with and developing technology, the technologist's decisions and actions are understood as part of a narrative history, a story. Action becomes intelligible only within a story in which "[w]e place the agent's intentions ... in causal and temporal order with reference to their role in his or her history." [54] Just as is the case for technologists, the decisions and actions of clients, consumers, and other stakeholders can be understood as part of a narrative, and are thus only intelligible by reference to this history [55]. Rights and individualist notions of autonomy are not sufficient to sustain a caring and empathetic practice [56]. Furthermore, attention to the stakeholder's narrative history in its full context can help us better realize Kantian values, such as autonomy [57].

Because understanding and committing to the goods internal to the practice of making, developing, and using technology requires one to take up the practice and engage with others who are involved in the practice and to contemplate the historical basis of current practice, practitioner's ethical character should be nurtured in order that they become virtuous technologists, who are inclined to act as they should [58]. Cultivating relational virtues, such as openness and responsiveness allows the technologist to realize the goods internal to the caring practice of science and technology, which include attending to and prioritizing the needs of the stakeholder as those needs become apparent [59].

Related to the consideration of narrative is the challenge of engineering practice in a situation of cultural and value diversity. Culturally based values and views have important implications for human well being and professional practice in terms of how professionals, clients, consumers, and other stakeholders think and behave, and in terms of what they expect [60]. We know that in engineering and technology, there are important differences in culturally based values and views between clients and customers and the technologists from whom they seek assistance or expertise [61]. How technology professionals think about and approach these issues has very real consequences for customers, clients, and communities, because if cultural differences are ignored or poorly addressed, they can result in conflict, decreased satisfaction, and problems with analysis, development, distribution and compliance, and even contribute to social inequalities [62]. We have also the example from architecture of how the design and construction of buildings helped shape the war in Syria [63]. We can do better. Evidence indicates that approaching cultural differences with sensitivity and skill can lead to better understanding [64] and better outcomes [65].

The goal of scientific and technological development is helping people and making the world better in some respect [66], but what that means is not always immediately evident. Consider an example from medicine. Suppose a patient refuses recommended care, be it evaluation or treatment [67]. Such cases bring into focus the complex relationship between respecting autonomy and fulfilling one's duty to advance the health of the patient. A physician might think treatment is the best course of action, and would select it, whereas the patient might refuse for reasons, such as religious belief, which the physician does not share. In approaching these decisions, it is necessary to be attentive to the specific beliefs and cultural backgrounds of patients [68].

Some authors have proposed "cultural humility" as an essential quality of scientists and technologists practicing in a multicultural world [69]. Cultural humility begins with self-awareness, self-reflection, and self-critique. It requires an open and respectful attitude toward diversity, and it recognizes the legitimacy of alternative ways of thinking and being. Cultural humility involves a willingness to learn about the unique perspectives of individual clients, consumers, and partners and the communities they come from; it also allows for the possibility that technicians themselves might grow and change as a result of appreciating others in this way. This means that engineers and other technicians can develop ethical sensitivity by cultivating a deep respect for others as partners in complex, dynamic relationships [70].

8.7 Communication and Ethics

Good engineering and technical practice requires good communication. In fact, good communication is not separable from good ethics. Ethical values are transmitted through explicit and implicit communication, and clear communication respects many ethical norms, including valuing truth-telling and respect for the listener. Ethical values may be explicitly and clearly communicated, what scholars call "espoused values." Alternatively, they may be implicit and communicated through practices, actions, habits, policies, and so on [71].

One unavoidable characteristic of working with emerging technologies is the experimental nature of the work. We often have some level of uncertainty about materials, outcomes, uses, likely benefits, and possible risks [72]. For these reasons, the development and deployment of technologies, especially new technologies, can usefully be thought of as an experimental process. Thinking about engineering and technical work in this way highlights some of the ways that ethical values and communication intersect.

Engineering introduces new products and processes that change—sometimes slightly and sometimes quite significantly—individual lives, patterns of life and work, ways of interacting (Facebook and online dating, for instance), economies and political institutions, and the natural world. This is done in situations of significant uncertainty about process and outcomes, and typically without much feedback (except through market mechanisms) from people whose lives will be changed.

Engineers should seek to provide good, clear, and full information about new processes and products and strive to bring affected parties into the conversation as early as possible. This requires managerial buy-in and, in some instances, placing ethical obligations over proprietary interests. Thus, engineers have great responsibility to stay broadly informed about the history of engineering and similar projects, about current contexts and the implications of their work, and about public feedback. The obligation to provide information places a *positive duty*, a duty not only to avoid harm or error, but also to take action to promote an ethically good outcome. In this case, the positive duty on engineers is to communicate actively about their work, beyond packaging inserts, user manuals, or product labels.

Espoused values can be defined as "the values a person or organization expresses or publishes in some manner" [73]. The espoused values of an organization are often articulated in various documents, goal and value statements, and policies. One of the best places to look for the explicit espoused values of any organization is the mission or goal statement. Development and dissemination of these documents is a standard and widespread practice, so much so that the work that these texts perform is at risk of being overlooked. These statements serve multiple functions. They can be used to motivate and inspire, as in Patagonia's mission statement, "Build the best product, cause no unnecessary harm, use business to inspire and implement solutions to the environmental crisis." They can also telegraph expectations and provide cues regarding current/future desired behavior, as in Google's post-2015 statement that the organization's mission is to "organize the world's information and make it universally accessible and useful." Interestingly, this statement was a significant revision from the values espoused in Google's pre-2015 motto "don't be evil."

Organizations espouse values in documents other than mission or goal statements. Ethics statements and codes of conduct are also important instances of written articulations of the ethical expectations and values of an organization and provide guidance for its members/employees [74]. Although over 75 % of US-based businesses now have codes of conduct/ethics [75], they are a relatively recent phenomenon, with most being written since 1970 [76]. These statements speak simultaneously to internal and external audiences. Internal to the organization, they play both instrumental and constitutive roles by articulating the ethical values and norms expected for individuals and for the organization. Externally, they help those who might interact with the organization to know what sorts of behavior it is reasonable to expect. Ethics statements also serve a publicity function and a legitimating function insofar as they communicate the ethical aspirations and values of an organization [77]. Some scholars view codes of conduct/ethics cynically, arguing that these statements primarily serve to coerce and/or control organizational members, while enhancing and/or protecting the organization's external status [78].

Ethical values are communicated in what an organization or individual says and in what an organization does. Values are enacted and communicated through actions taken. These actions might be consistent with expressed values, as when Staples worked with HP to reduce e-waste from all sources, in keeping with their increased corporate emphasis on sustainability. Another example is the recognition of Marriott in 2016, for the tenth straight year, as one of the world's most ethical companies by the Ethisphere Institute in keeping with it's claim that "How we do business is as important as the business we do" [79]. Conversely, actions might not be consistent with espoused claims, as when the 2010 explosion of BP's Deep Water Horizon oil drilling platform and subsequent oil spill in the Gulf of Mexico stood in stark contradiction to BP's explicit goal of being an innovative and environmentally friendly energy company [80].

The actions taken have both material and symbolic impact. The material impact of the action can support or contradict the relevant explicit value. The action taken communicates either that the organization or person is a consistent actor or that, in reality, the person or organization follows a quite different value from that stated. Some argue that the values implicit in the actions taken are the actual values, and the explicit but not enacted values are thus articulations of self-deception or bad faith [81]. According to this view, whatever an organization or individual does, the impact of the actions should be the focus of analysis for identifying the values of the organization or individual.

Ethical values are also enacted and communicated in practice. A practice is comprised of shared understandings, rules, and acceptable ends that are inherently tied to context, such that all factors are mutually defined through the relationships [82]. Put differently, ethical values may emerge in a dynamic performance of saying and doing [83]. For example, there are 759 full members of the Association for the Advancement of Sustainability in Higher Education (AASHE). One of the strategies advocated for reducing energy consumption and the carbon footprint on residential campuses was to supply drying racks in residence halls. Doing so is a practice likely to be consistent with espoused sustainability goals at member colleges and universities. However, none of those factors (values statements, membership in a relevant organization, and material practices), singly or together, can ensure that students use the drying racks. For humans to act in a particular way, especially if those actions run counter to existing norms of behavior or demand more emotional, psychological, mental, or moral effort than other options, requires either a particular individual character or an organizational culture that supports the action in question. As we know, at least since Aristotle, the actions of individuals and the character of the culture they comprise are inextricable intertwined.

Finally, talking about ethics—about values, expectations, and behaviors—encourages ethical sensitivity and helps develop ethical skills. Considerable research suggests that courses in ethics help in raising awareness of ethical issues and encourage more responsible and critical thinking, and better action [84].

Two of the great challenges of emerging technologies are (i) creating organizations, workplaces, corporations, professional societies, agencies, and schools that encourage clear and timely communication and high ethical standards; and (ii) nurturing and supporting individual practitioners who work according to good principles of ethics and communication.

8.8 Final Remarks

Ethical theories provide a framework to help us determine what is right and wrong, good and bad. They provide a set of standards that allows us to evaluate particular actions or the characters of people. Although the theories presented here are competing ethical theories, and although a critical evaluation of their tenets can bring to the forefront some serious concerns about each theory, they can be helpful in guiding professional practice. As mentioned earlier, the usefulness and importance of these approaches is not predicated on a simpleminded adherence to any one theory in a professional situation. However, together with awareness of the principles of good communication practices, ethical theories can help us reflect on what is important and valuable [85], communicate with others [86], and think about how best to achieve the ends of science and technology.

8.9 Questions

1. Define "ethical sensitivity." Why is it important? Give an example of when ethical sensitivity is important in the development, manufacture, or use of new technologies.

2. Define "ethical motivation." Why is it important? Give an example of when ethical motivation is important in the development, manufacture, or use of new technologies.
3. Define "ethical judgment." Why is it important? Give an example of when ethical judgment is important in the development, manufacture, or use of new technologies.
4. Define "ethical character." Why is it important? Give an example of when ethical character is important in the development, manufacture, or use of new technologies.
5. What are "espoused values?" How are they important to ethical communication?
6. What does it mean to say "goods are internal to a practice?" Give an example from technology or engineering.
7. Which ethical framework is most concerned with maximizing outcomes, or consequences?
8. Which ethical framework is most concerned with ethical character?
9. Which ethical framework is best for evaluating ethical motivation?
10. Why are technical knowledge and skill alone insufficient to be an ethically responsible technologist?
11. Explain why new technologies create new ethical concerns and challenges.

Bibliography

[1] Hippocrates. Oath of Hippocrates (Fourth Century B.C.E.). In: Reich WT (ed). Encyclopedia of Bioethics (vol. 5). Macmillan, p. 2632. 1995.
[2] Proctor RN. Racial hygiene: Medicine under the Nazis, Harvard University Press, 1988.
[3] Reverby SM, ed. Tuskegee's truths: Rethinking the Tuskegee syphilis study, University of North Carolina Press, 2000.
[4] Pfrang EO, Marshall R. Collapse of the Kansas City Hyatt Regency walkways. Civil Engineering – ASCE, 1982, 52(7), 65–69.
[5] Boisjoly R. Ethical decisions: Morton Thiokol and the space shuttle Challenger disaster. American Society of Mechanical Engineers Winter Annual Meeting, Boston, Massachusetts, (pp. 1–13), 1987, December.
[6] Carson R. Silent spring, Houghton Mifflin Harcourt, 2002.
[7] Braunack-Mayer A. What makes a good GP? An empirical perspective on virtue in general practice. Journal of Medical Ethics, 2005, 31, 82–87; Singer PA, Pellegrino ED, Siegler M. Clinical ethics revisited. BMC Med Ethics, 2001, 2, E1; Morenz B, Sales B. Complexity of Ethical decision Making in Psychiatry. Ethics and Behavior, 1997, 7(1), 1–14.
[8] Sass HM. The clinic as testing ground for moral theory: A European view. Kennedy Institute of Ethics Journal, 1996, 6(4), 351–355.
[9] Baker R, McCullough LB. Medical Ethics' Appropriation of Moral Philosophy: The Case of the Sympathetic and the unsympathetic physician. Kennedy Institute of Ethics Journal, 2007, 17(1), 3–22.
[10] Ortega y Gasset J. History as a system, W.W. Norton and Company, 1941.
[11] Fraser N. Talking about needs: Interpretive contests as political conflicts in welfare-state societies. Ethics, 1989, 99 (Jan. 1989), 291–313.

[12] Mumford L. Technics and civilization, Harcourt, Brace & Company, Inc., New York, 1934; Ortega y Gasset J. History as a system, New York, W.W. Norton and Company, 1941; Ellul J. The Technological Society, New York, Knopf, 1964; Jonas H. Toward a philosophy of technology. Hastings Center Report, February 1979, pp. 34–43; Jonas H. The imperative of responsibility: In search of ethics for the technological age, Chicago, University of Chicago Press, 1985.

[13] Ziman JM. Why must scientists become more ethically sensitive than they used to be? Science, 1998, 282, 1813–1814.

[14] Anderson MS. Normative orientations of university faculty and doctoral students. Science and Engineering Ethics, 2000, 6(4), 443–461.

[15] Thompson PB. Food biotechnology in ethical perspective, Blackie Academic & Professional/Chapman & Hall, 1997; Yount L. Biotechnology and genetic engineering, Facts On File, Inc., 2000; Sandel MJ. The case against perfection: Ethics in an age of genetic engineering, Belknap Press of Harvard University, 2009.

[16] Martin M, Schinzinger R. Ethics in engineering, 2nd edn., McGraw-Hill Book Company, 1989; Moriarty G. The engineering project: Its nature, ethics, and promise, The Pennsylvania State University Press, 2008.

[17] ABET. Criteria for accrediting engineering programs, 2012–2013, 2012. Retrieved from www.abet.org/engineering-criteria-2012-2013/.

[18] Harris CE Jr. Engineering ethics: From preventative ethics to aspirational ethics. In: Michelfelder DP et al. (eds). Philosophy and engineering: reflections on practice, principles, and process, Springer, Dordrecht, Netherlands, 2013.

[19] Center for Engineering Ethics and Society. Overcoming challenges to infusing ethics into the development of engineers workshop. National Academy of Engineering, 2017.

[20] https://www.nspe.org/resources/licensure/state-ce-requirements.

[21] Koen BV. Discussion of the method: Conducting the engineer's approach to problem solving, Oxford University Press, 2003.

[22] O'Fallon MJ. A review of the empirical ethical decision-making literature: 1996–2003. Journal of Business Ethics, 2005, 59, 375–413.

[23] Rest J. Moral development: Advances in research and theory, Praeger, New York, 1986.

[24] Solomon RC. Ethics and excellence: Cooperation and integrity in business. Oxford University Press, New York, 1993.

[25] Hanks EK, Hanks JC. Rules, compliance, and ethics – oh my, The Public Manager, September 2015.

[26] Ortega y Gasset J. Concord and liberty, W. W. Norton and Company, 1946.

[27] Habermas J. Between facts and norms: Contributions to a discourse theory of law and democracy, The MIT Press, Cambridge, MA, 1998.

[28] Bayles M. Professional ethics, Wadsworth Publishing, 1988; Israel M, Hay I. Research ethics for social scientists: between ethical conduct and regulatory compliance, Sage Publications, 2006; Kultgen J. Evaluating codes of professional ethics. In: Robinson WL, Pritchard MS (eds). Profits and professions. Essays in business and professional ethics, Humana Press, pp. 225–263, 1983; Martin M, Schinzinger R. Ethics in engineering, 2nd edition, McGraw-Hill Book Company, 1989; Regan T. Research ethics: An introduction, Ethics in Science and Engineering National Clearinghouse, Paper 293, 2002. Accessed on 16 October, 2013 at http://scholarworks.umass.edu/esence/293; Whitman ME, Mattford HJ. Principles of information security, 4th Edition, Cengage Learning, 2012.

[29] McDonald GM, Donleavy GD. Objections to the teaching of business ethics. Journal of Business Ethics, 1995, 14(10), 839–853.

[30] Moriarty G. The engineering project: Its nature, ethics, and promise, The Pennsylvania State University Press, 2008; Mumford L. Technics and Civilization, Harcourt, Brace & Company, Inc., 1934.

[31] Woodruff P. Living toward virtue: Practical ethics in the spirit of Socrates. Oxford University Press, New York, 2022.

[32] Widen WH. Enron at the Margin. The Business Lawyer (pp. 961–1002), 2003.

[33] Fitts WT, Ravdin IS. What Philadelphia physicians tell patients with cancer. JAMA, 1953, 153, 901–904.

[34] Oken D. What to tell cancer patients: a study of medical attitudes. JAMA, 1961, 175, 1120–1128.

[35] Heubel F, Biller-Andornol N. The contribution of Kantian moral theory to contemporary medical ethics: a critical analysis. Med Health Care Philos, 2005, 8(1), 5–18.

[36] http://www.acm.org/about/se-code#full.

[37] Martin M, Schinzinger R. Ethics in engineering, 2nd edn., McGraw-Hill Book Company, 1989.

[38] Forrow L, Wartman SA, Brock DW. Science, Ethics, and the making of clinical decisions. JAMA, 1988, 259, 3161–3167.

[39] Trials of war criminals before the Nuernberg Military Tribunals under control council law 2(10), pp. 181–182; Washington, D.C., U.S. Government Printing Office, 1949. Available at https://www.loc.gov/item/2011525364/, accessed 5 October 2023.

[40] World Medical Association Declaration of Helsinki. Ethical principles for medical research involving human subjects. http://www.wma.net/e/policy/b3.htm, accessed 19 April 2007.

[41] Beauchamp TL, Childress JF. Principles of biomedical ethics, 5th edn., Oxford University Press, 2001.

[42] Novak DH, Plumer R, Smith RL et al. Changes in physicians' attitudes toward telling the cancer patient. JAMA, 1979, 241, 897–900.

[43] Veatch RM. Disrupted dialogue, Oxford University Press, New York, 2005; Ramsey P. The patient as person, New Haven, CT, Yale University Press, 1970.

[44] Faden RR, Beauchamp TL. A history and theory of informed consent, Oxford University Press, 1986; Beauchamp TL, Childress JF. Principles of biomedical ethics, 5th edn., Oxford University Press, 2001.

[45] Mill JS. Utilitarianism, Dolphin Books, 1961.

[46] Singer P. Practical ethics, 2nd edn., Cambridge University Press, 1993.

[47] Jones C. The utilitarian argument for medical confidentiality: a pilot study of patients' views. Journal of Medical Ethics, 2003, 29, 348–352.

[48] Pellegrino ED. Toward a virtue-based normative ethics for the health professions. Kennedy Institute of Ethics Journal, 1995, 5, 253–277; Benner P. A dialogue between virtue ethics and care ethics. Theor Med, 1997, 18, 47–61.

[49] Pellegrino ED, Thomasma DC. Virtues in medical practice, Oxford University Press, New York, 1993.

[50] Gregory J. John Gregory's writings on medical ethics and philosophy of medicine, McCullough LB (ed). Kluwer Academic Publishers, 1998; More E. "Empathy" enters the profession of medicine. In: More E, Milligan M (eds). The Empathic Practitioner, Rutgers University Press, pp. 19–39, 1994.

[51] Aristotle. Nicomachean ethics, trans by Terence Irwin Indianapolis, Hackett Publishing Company, 1985.

[52] MacIntyre A. After virtue, 2nd edn., University of Notre Dame Press, p. 184, 1984.

[53] MacIntyre A. After virtue, 2nd edn., University of Notre Dame Press, p. 187, 1984.

[54] MacIntyre A. After virtue, 2nd edn., University of Notre Dame Press, p. 108, 1984.

[55] Brody H. Stories of sickness, 2nd edn., Oxford University Press, 2003; Charon R. Narrative medicine: a model for empathy, reflection, profession, and trust. JAMA, 2001, 286, 1897–1902; Jones AH. Literature and medicine: narrative ethics. Lancet, 1997, 349, 1243–1246; Hunter KM. Doctors' stories: The narrative structure of medical knowledge, Princeton University Press, 1991.

[56] Halpern J. From detached concern to empathy: humanizing medical practice, Oxford University Press, 2001.

[57] Schafer C, Putnik K, Dietl B et al. Medical decision-making of the patient in the context of the family: results of a survey. Support Care Cancer, 2006, 14(9), 952–959; Slowther A-M. The role of the family in patient care. Clinical Ethics, 2006, 1(4), 191–193; Tauber AI. Patient autonomy and the ethics of responsibility, The MIT Press, 2005.

[58] Rhodes R, Cohen DS. Understanding, being and doing: Medical ethics in medical education. Cambridge Quarterly of Healthcare Ethics, 2003, 12(1), 39–53.

[59] Benner P. A dialogue between virtue ethics and care ethics. Theor Med, 1997, 18, 47–61.

[60] English-Lueck JA, Darrah CN, Saveri A. Trusting strangers: work relationships in four high-tech communities. Information, Communication & Society, 2002, 5(1), 90–108; Allchin D. Values in science: An educational perspective. Science education and culture, Springer Netherlands, 2001, pp. 185–196; Perkins HS, Geppert CMA, Gonzales A et al. Cross-cultural similarities and differences in attitudes about advance care planning. J Gen Intern Med, 2002, 17, 48–57; Perez-Stable E, Sabogal F, Otero-Sabogal R et al. Misconceptions about cancer among Latinos and Anglos. JAMA, 1992, 268, 3219–3223.

[61] Blackhall LJ, Murphy ST, Frank G. Ethnicity and attitudes towards patient autonomy. JAMA, 1995, 274, 820–825; 274, 826–829; Pachter LM. Culture and clinical care: folk illness beliefs and behaviors and their implications for health care delivery. JAMA, 1994, 271, 690–694.

[62] Barker JC. Cultural diversity: changing the context of medical practice. Western Journal of Medicine, 1992, 157, 248–254; Galanti G. Caring for patients from different cultures, case studies from American hospitals, University of Pennsylvania Press, 1992.

[63] Al-Sabouni S. The battle for home: The vision of a young architect in Syria, Thams and Hudson, 2016.

[64] Carrese JA, Perkins HS. Ethics consultation in a culturally diverse society. Public Affairs Quarterly, 2003, 17, 97–120; Carrese JA, Rhodes LA. Bridging cultural differences in medical practice: the case of discussing negative information with Navajo patients. J Gen Intern Med, 2000, 15, 92–96.

[65] Shain RN, Piper JM, Newton ER et al. A randomized, controlled trial of a behavioral intervention to prevent sexually transmitted disease among minority women. The New England Journal of Medicine, 1999, 340, 93–100; Tervalon M, Murray-Garcia J. Cultural humility versus cultural competence: a critical distinction in defining physician training outcomes in multicultural education. J Health Care Poor Underserved, 1998, 9, 117–125.

[66] Nowotny H, Scott PB, Gibbons MT. Re-thinking science: Knowledge and the public in an age of uncertainty, John Wiley & Sons, 2013; Kitcher P. Science in a democratic society. Scientific realism and democratic society, Brill, 95–112, 2011; Kitcher P. Science, truth, and democracy, Oxford University Press, 2003; Pellegrino ED, Thomasma DC. For the patient's good: the restoration of beneficence in health care, Oxford University Press, 1988.

[67] Carrese JA. Refusal of care: patients' well-being and physicians' ethical obligations; "but doctor, I want to go home." JAMA, 2006, 296, 691–695.

[68] Akabayashi A, Slingsby BT. Informed consent revisited: Japan and the U.S. The American Journal of Bioethics, 2006, 6(1), 9–14; Gold M. Is honesty always the best policy? Ethical aspects of truth-telling. Internal Medicine Journal, 2004, 34, 578–580.

[69] Groll L. Negotiating cultural humility: First-year engineering students' development in a life-long journey, ProQuest LLC, 2013; Allenby BR. Engineering and ethics for an anthropogenic planet. National Academy of Engineering, 2004, pp. 9–28; Hunt LM. Beyond cultural competence: applying humility to clinical settings. Bull Park Ridge Cent, 2001, Nov–Dec, pp. 3–4; Tervalon M, Murray-Garcia J. Cultural humility versus cultural competence: a critical distinction in defining physician training outcomes in multicultural education. J Health Care Poor Underserved, 1998, 9, 117–125.

[70] Akabayashi A, Slingsby BT. Informed consent revisited: Japan and the U.S. The American Journal of Bioethics, 2006, 6(1), 9–14; Slingsby BT. The nature of relative subjectivity: A philosophical perspective on Japanese thought. The Journal of Medicine and Philosophy, 2005, 30, 11–29.

[71] Hanks EK, Hanks JC. Organizational values. In: Farazmand A (ed). Global Encyclopedia of Public Administration, Public Policy, and Governance, Springer, 2016.

[72] Feenberg A. Between reason and experience: Essays in technology and modernity, MIT Press, 2010; Martin M, Schinzinger R. Ethics in Engineering, McGraw-Hill, 1983.

[73] Gehman J, Trevino LK, Garud R. Values work: A process study of the emergence and performance of organizational values practices. Academy of Management Journal, 2013, 56(1), 84–112.

[74] Schwartz M. The nature of the relationship between corporate codes of ethics and behavior. Journal of Business Ethics, 2001, 32(3), 247–262.

[75] Somers MJ. Ethical codes of conduct and organizational context: A study of the relationship between codes of conduct, employee behavior and organizational values. Journal of Business Ethics, 2001, 30(2), 185–195.

[76] Berenbeim R. Corporate ethics: Research report #900, The Conference Board, 1987.

[77] Dowling J, Pfeffer J. Organizational legitimacy: Social values and organizational behavior. Pacific sociological review, 1975, pp. 122–136; Kultgen JH, Ethics and professionalism, University of Pennsylvania Press, 1988; Levy DL, Kolk A. Strategic responses to global climate change: Conflicting pressures on multinationals in the oil industry. Business and Politics, 2002, 4(3), 275–300; Valentine S, Barnett T. Ethics code awareness, perceived ethical values, and organizational commitment. Journal of personal selling & Sales Management, 2003, 23(4), 359–367.

[78] Kultgen J, Alexander-Smith R. The ideological use of professional codes [with Commentary]. Business & Professional Ethics Journal, 1982, 1(3), 53–73.

[79] Ethisphere Institute. The 2016 World's Most Ethical Companies. Ethisphere Magazine, 2016, 10, 1; Marriott Corporation. Core Values and Heritage, accessed online on 31 July 2016 at http://www.marriott.com/culture-and-values/core-values.mi.

[80] Harlow WF, Brantley BC, Harlow RM. BP initial image repair strategies after the Deepwater Horizon spill. Public Relations Review, 2011, 37(1), 80–83; Ingersoll C, Locke RM, Reavis C. BP and the Deepwater Horizon Disaster of 2010. MIT Sloan Management Review, 2012, 10(110), 7.

[81] De Beauvoir S. The ethics of ambiguity, Citadel Press, 1948; Camus A, The Myth of Sisyphus and Other Essays, Justin O'Brien (trans), Vintage, 1955; Sartre JP, Jean-Paul Sartre: Basic Writings, Routledge, 2002.

[82] Shapiro B, Naughton M. The expression of espoused humanizing values in organizational practice: A conceptual framework and case study. Journal of Business Ethics, 2015, 126(1), 65–81.

[83] Jonsen K, Galunic C, Braga T. Evaluating espoused values: Does articulating pay off? European Management Journal, 2015, 33, 332–340, 333.

[84] Cohen E, Cornwell L. A question of ethics: Developing information system ethics. Journal of Management Development, 1989, 2(2), 28–38; Fritzsche DJ, Becker H. Ethical behavior of marketing managers. Journal of Business Ethics, 1983, 2(4), 291–299; Gautschi FH, Jones TM. Enhancing the ability of business students to recognize ethical issues: An empirical assessment of the effectiveness of a course in business ethics. Journal of Business Ethics, 1998, 17(2), 205–216; Lind G. The theory of moral-cognitive development: A socio-psychological assessment. In: Lind GHA, Wakenhut R. (eds). Moral judgments and social education (pp. 21–54), New Brunswick, Transaction, 2010; Martin TR. Do courses in ethics improve the ethical judgment of students? Business Society, 1981, 20(2), 17–26; McDonald GM, Donleavy GD. Objections to the teaching of business ethics. Journal of Business Ethics, 1995, 14(10), 839–853; Stead BA, Miller JJ. Can social awareness be increased through business school curricula? Journal of Business Ethics, 1988, 7(7), 553–560; Walker M. Evaluating the intervention of an ethics' class in students' ethical decision-making: A summative review. International Education Studies, 2013, 6(1), 10–25.

[85] MacIntyre A. After virtue, 2nd edn., University of Notre Dame Press, 1984.

[86] Irish R, Weiss PE. Engineering communication: from principles to practice, Oxford University Press, 2013.

Christie M. Sayes, James Y. Liu, and Matthew Gibb

9 Behavior-Based Worker Safety for Engineered Nanomaterials

9.1 Introduction

Incorporating engineered nanomaterials, such as dry powders, composites, adhered films, and particle suspensions in solvents, can provide distinct benefits for products [1]. The manufacture of these novel products uses techniques such as lithography and printing, milling and grinding, epitaxy, self-assembly, sol-gel, and flame pyrolysis [2]. Nanopowder refers to specially crafted particles less than 100 nanometers in diameter. By convention, the dimensions of all synthesized, produced, and manufactured nanopowders are defined as diameters less than 100 nanometers in at least one dimension. Some unique advantages of so-called "nano-enabled products" include materials effectively made to be stronger, lighter, more durable, more reactive, more sieve-like, or better electrical conductors, among many other traits (NNI, http://www.nano.gov/you/nanotechnology-benefits). However, the attractive benefits of nanomaterials are also associated with some significant risks to occupational workers who function in various stages of the nano-enabled product life or life cycle [3, 4].

Following specific safety protocols when dealing with dry powders is essential due to their high surface-to-volume ratio and reactivity. Powder explosions can lead to severe harm, and even fatalities [5]. Exposure to airborne particles has induced adverse effects in human populations. Pulmonary (i. e., lung) effects are especially worrisome when particles are inhaled chronically [6]. Though exposure to composites, adhered films, and particle suspensions is generally considered less risky for occupational exposure, it's important to note that dermal effects can still occur if there is close contact with these types of nanoparticle systems [7–10].

The extent of toxicological effects has yet to be fully understood for each engineered nanomaterial being studied today [11, 12]. To complicate matters further, the resulting adverse pathology is sometimes observed after some time. Because of this, illnesses may develop years after the initial exposure, making it very difficult to determine a cause-effect relationship between exposure and disease onset. This book chapter proposes some guidelines for safely handling engineered nanomaterials and nano-enabled products. These guidelines can aid in preventing exposure and developing exposure-related illnesses in the occupational setting. Improved behavior-based worker safety for engineered nanomaterials can be achieved with increased mindfulness of the risks associated with dry powders, composites or adhered films, and particle suspensions in solvents.

https://doi.org/10.1515/9783110781830-009

9.2 Traditional Behavior-Based Worker Safety

Nanotechnology applies science and behavior to real-world problems, because research and development are needed to launch a product or process enabled by the nanomaterials [13]. However, the various risks to workers developing these technologies must be met with adequate and evolving safety measures and regulations [3]. One of these safety regulations is a program known as Behavior-Based Safety. Behavior-Based Safety (BBS) is conventionally defined as "the process that creates a safety partnership between management and employees that continually focuses people's attentions and actions on their daily safety behavior" [14]. This definition applies to the nanotechnology industry in that communication among workers at the research and development facilities and a comprehensive understanding of the technology are critical to safety in the workplace.

Developing a behavior plan should include company culture, employees' past experiences, and workers' physiological states. In nanoscience research, workers will inevitably be exposed to engineered nanomaterials, such as aerosols or suspensions. Therefore, companies should provide adequate personal protective equipment (PPE) and engineering controls to mitigate exposure and decrease health risks. Furthermore, nanoparticles have been found to exacerbate an already unhealthy individual [15]. If a person has chronic lung disease, such as Chronic Obstructive Pulmonary Disease (COPD), inhaling nanoparticles could exacerbate and increase the amount of damaged tissue, exacerbating allergic reactions or diminishing defensive immune responses [16].

In the field of material science and engineering, most industries follow a set of established guidelines for promoting worker safety. [17]. Occupational health practitioners in each workplace typically complete six tasks to educate, assess, and enhance safety measures:
(1) Identify hazardous issues within the workspace as a whole;
(2) Identify hazardous issues associated with an individual's task;
(3) Create a forum to engage in open conversations to educate and discuss hazardous issues;
(4) Draft a priority list of improvements concerning time and resources;
(5) Assess the results of the implemented improvements;
(6) Advocate for good housekeeping in the working environment.

When applying these tasks to a nanomaterial-specific workplace, the basic practices remain the same, but a few additional considerations should be added:
(1) Involve all parties in the workplace, including the administration;
(2) Form an advisory committee that is knowledgeable of the nanomaterial literature;
(3) Maintain proper labeling, packaging, and marking of each nanomaterial component;
(4) Increased training in spontaneous combustion situations;
(5) Increased use of personal protective equipment, including frequently changing gloves and other body coverings.

9.3 The ABC Model as Applied to Nanotechnology in the Workplace

One specific behavior-based safety program type is the "ABC" model [18, 19]. ABC is the acronym of Antecedent-Behavior-Consequence, where "A" stands for Antecedent: a thing or event that existed before or logically precedes another; "B" stands for Behavior: how a person acts in response to a particular situation or stimulus, and "C" stands for Consequence: a result or effect of an action or condition. The ABC model is designed to collect data and allow for workplace safety analyses, i. e., an assessment tool used to gather the information that should evolve into a positive behavior support plan (Figure 9.1).

THE OEHS FRAMEWORK

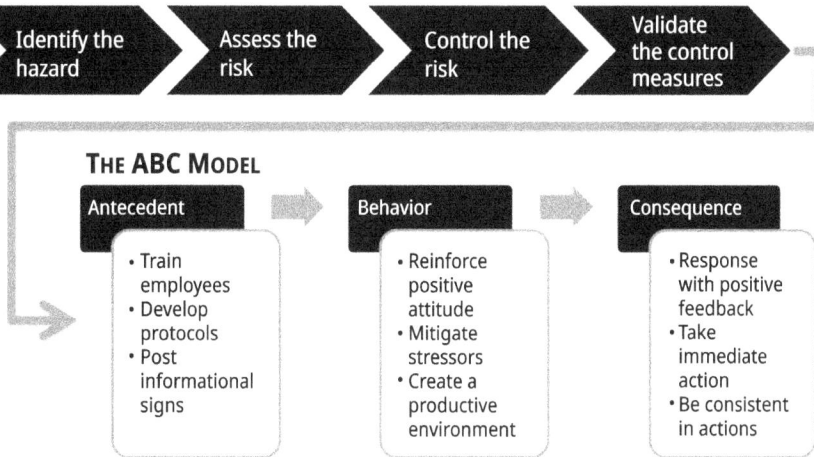

Figure 9.1: The flow of information from the occupational and environmental health and safety (OEHS) framework to the antecedent, behavior, consequence (ABC) model.

In nanotechnology, identifying potential hazards can be difficult. Nanomaterials are so small that they cannot be seen with the naked eye, making detecting spills or leaks in a production system challenging. To stay safe, it is recommended to prioritize the following steps: (1) designing out hazards, (2) implementing engineering controls to minimize risk, (3) training employees on safe behavior, and (4) using personal protective equipment (PPE) [20].

Promoting safe behavior can lead to increased business success and boost employee morale. However, failure to prioritize safety may result in consequences such as regulatory intervention and a higher employee turnover rate due to increased risk of injury or illness resulting from insufficient knowledge of the unique technology and its proper operation. Companies need to tailor behavior-based safety principles to fit their specific needs, including those in emerging industries, such as nanotechnology, to improve or-ganizational performance and meet market demands.

9.4 Exposure Scenarios Along the Nanomaterial Value Chain

Depending on their hazardous nature, nanomaterials can pose risks to occupational workers, consumers, and the environment. Even waste products and intermediate materials can be potentially harmful. An environmental risk assessment is often conducted along the value chain to assess the risks associated with these materials. Environmental risk analysis aims to understand the events and activities that may risk human health or the environment [21–23].

Exposure to nanoparticles can occur at any point along the product value chain. There are unique stages to developing and implementing nano-enabled products, from research and development to implementation and use to removal and recycling. The four stages of a product value chain are production and manufacturing, distribution and transit, formulation and use, and disposal and recycling. In each stage of the custody chain, different people are exposed to varying forms of nanoparticles for different amounts of time. Understanding the exposure in each stage and the possible health outcomes after each unique exposure can lead to a safer work environment.

The body can be exposed to nanomaterials through four main routes: dermal and eye contact, inhalation, and ingestion. The health effects of each of these routes are discussed in the following section.

1. Dermal exposure (e. g., skin contact) may lead to irritation, sensitization, or allergic reaction. Nanopowders have been shown to penetrate the epidermal layers and intercalate within cells [24].
2. Ocular exposure (e. g., eye contact) can result in irritation and permanent damage. Nanopowders have been shown to absorb through the mucous layers of the eyeball and into the tear ducts [25, 26].
3. Inhalation exposure (e. g., breathing) can result in both lung structural and functional changes. In the case of inhalation, the size of the particles determines the level of penetration into the lungs. The smaller particles can penetrate deeper into the lung and cause more damage to the organ. Continuous damage and inflammation can lead to fibrosis, cancer, and other chronic diseases. For example, titania (TiO_2) nanopowders have been shown to decrease lung function and cause fibrosis of the lung tissues [27].
4. Ingestion exposure (e. g., swallowing) can alter the commensal microbial environment of the gastrointestinal tract. Nanopowders have demonstrated antimicrobial effects in bacteria associated with normal intestinal digestion [28].

Once exposed, nanoparticles of a small enough size can enter through epithelial barriers in various tissues and travel through the bloodstream to distribute throughout the body. For instance, from the gastrointestinal system nanoparticles can be transported into the circulatory system and to subsequent organs (e. g., the brain) [29].

The nanotechnology industry often discusses the value chain, which refers to adding value to a nano-enabled product from discovery to end-of-life [30–32]. This process is closely linked to the product's life cycle analysis, which examines its lifespan from raw materials to the final disposal [33]. This analysis considers the potential risks and impacts on workers, consumers, and the environment [34]. The nanomaterial product value chain comprises four stages: production and manufacturing, distribution and transit, formulation and end-users, and disposal and recycling (Figure 9.2).

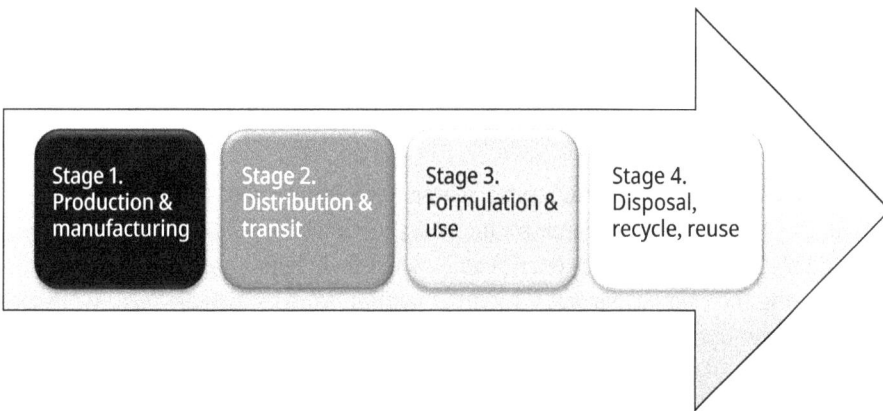

| Stage 1. Production & manufacturing | Stage 2. Distribution & transit | Stage 3. Formulation & use | Stage 4. Disposal, recycle, reuse |

Figure 9.2: The figure describes the different stages of the product value chain for nano-enabled products, materials, and particles. Stage 1 involves producing or manufacturing nanomaterials for the final product. Stage 2 deals with transporting nano-intermediates or materials to different locations. Stage 3 consists in formulating and introducing nano-enabled products to the market. Finally, Stage 4 focuses on the disposal, incineration, recycling, or reusing of the product or waste at the end of its life.

9.4.1 Stage 1: Production and Manufacturing

The production of nanoparticles poses the highest risk of exposure in the value chain [35, 36]. However, measures are available to minimize this risk in a laboratory or industrial setting. These measures include engineering controls and safety protocols, from safety training to using personal protective equipment and installing equipment such as fume hoods and ventilation systems (Figure 9.3).

Understanding the parameters within the production and manufacture of nanomaterials can help focus safety strategies. For example, during "top-down" synthesis, ultra-fine particles are generated through mechanical breakdown. Spontaneous combustion can occur if the particles are collected and stored in a single location, because small

Equipment Control

• Total enclosure of the process
• Partial enclosure and local exhaust ventilation
• Local exhaust ventilation
• Proper equipment

Personal Habits

• Limitation of number of workers
• Reduction in periods of exposure
• Use of Personal Protection Equiptment
• No eating or drinking in contaminated areas

Figure 9.3: Examples of employers implementing engineering control and personal protection within a nanotechnology workplace. A significant part of engineering controls involves hardware/software to maintain process parameters.

particles burn readily due to high surface area once aerosolized. This specific particle size range (i. e., <250 nm in diameter) increases the likelihood of a spontaneous combustion [37, 38]. Small particles burn readily when their ignition point is reached and tend to ignite the coarser particles. Potential ignition sources include heat due to mechanical stress, open flames, welding torches, matches and cigarettes, faulty electrical equipment, and static electric discharges. These conditions and behaviors must be minimized in work areas close to nanopowder production. To lessen workplace risks, it's recommended to take preventive measures, such as enclosing the system or installing proper ventilation. New technologies, such as sensors in storage areas to measure temperature or airborne concentrations, can also help mitigate workplace risks. In addition, in "bottom-up" synthesis of engineered nanomaterials, production involves the growth of nanoparticles in solvents or gas under high temperatures and pressures. This could lead to exposure to fumes, burns, or explosions near the occupational workers. Anticipating these potential risks can aid in creating a safe and healthy workplace environment.

9.4.1.1 Nanopowder Dust Generation, Collection, and Disposal

The generation of nanopowder dust poses a potential risk during both the production and manufacturing stages and the formulating and end-user stages in the product value chain [39, 40]. The National Institute for Occupational Safety and Health (NIOSH) recommends that worksites install dust collectors at the site of material manipulation. Fume hoods, safety cabinets, enclosures, and ductwork should be installed and constructed of rustproof, non-sparking metal. Differential pressure is a good way to prevent air-borne nanoparticles from leaving a designated area. Dust collection systems handling nanopowders should act as dedicated use systems. Periodically, the systems should be cleaned using a damp cloth and disposed of in sealed containers. Specific to fume hoods and safety cabinets, HEPA filters are strongly recommended as an effective barrier to preventing particles from appearing in recirculated air. Ductwork should be deliberate, because too little ductwork can lead to dangerous buildup.

9.4.1.2 Good Housekeeping Practices for Nanomaterials

Exposure to engineered nanomaterials is possible if one works in a production or manufacturing workroom, where they are used as starting materials [7, 40]. A dirty or unorganized workroom is more hazardous than a tidy one [41]. Therefore, good housekeeping is essential to prevent accidents and exposures in the workplace, especially within the nanopowder worksite. For instance, dust accumulations should not be allowed to concentrate on floors, piping, ductwork, conduit, or walls. Recommended cleaning tools include HEPA vacuum pickup and wet wiping methods. Other relevant examples of good housekeeping practices are not necessarily unique to nanomaterials but should be practiced nonetheless: avoid food or beverage consumption in workplaces where nanopowders are handled, require handwashing before and after each powder handling scenario, and provide facilities for showering and changing clothes. Dry sweeping or air hoses should never be used to clean nanopowder work areas, because these methods may aerosolize powders and increase worker exposure. Synthetic fiber brushes or plastic tools should be avoided, because these instruments tend to accumulate static charges, which, upon ignition, may cause a fire.

9.4.2 Stage 2: Distribution and Transportation

In the first stage of a product value chain, the risk of exposure primarily falls upon the employees and other personnel aware of the possible exposure. In Stage 1, the risk of exposure was contained to a laboratory or industrial setting. In Stage 2, the risk of exposure to nanomaterials may be unknown. The opportunity for nano-safety training for an occupational worker outside the nanomaterial laboratory or industrial setting may be less available or as comprehensive as needed. Occupational workers in the transportation sector will also handle engineered nanomaterials [42, 43]. These materials are presumably packaged but have the potential to spill or leak if mishandled. Therefore, inhalation, ingestion, ocular or dermal exposure is possible outside of the contained worksite.

Safe behavior practices are critical when nanomaterials are transported or delivered. The sender and recipient of the material should be well-informed and have primary responsibilities throughout the process. Safety measures include administrative, technical, or engineering facets and health and safety personnel. Oral and written communication also helps to ensure an understanding of the physical and chemical properties, the amount, and the distance traveled relevant to the specific nanopowder being transported.

Because an accident during the transport of dangerous goods can lead to catastrophic consequences, guidelines and regulations have been established to protect workers, society, and the environment (https://www.osha.gov/Publications/OSHA_FS-3634.pdf). Spillages are also possible when materials must be appropriately packaged, handled, or labeled; therefore, the material must be adequately prepared for shipment and storage. Finally, the risk of an accident increases when the material is left unattended.

9.4.2.1 Specific Conditions Can Increase the Risks

If certain conditions are not monitored, the distribution and transportation of nanomaterials can pose significant risks along the chain of custody [44]. For example, some engineered nanomaterials may become hazardous when the particles encounter air, reactive gases, water/humidity, or reactive solvents. Hygroscopic nanopowders could quickly generate heat. When some forms of iron nanoparticles react with water, for instance, reactive oxygen species are readily generated and can cause skin sensitization or other systemic toxicities. The pressure within sealed packages can rise in the heat and cause uncontrolled reactions (e. g., fire). By extension, any temperature change may affect the quality of the engineered nanomaterial, potentially rendering it useless.

In 2011, the United Nations published "Recommendations on the Transport of Dangerous Goods," a book written by a committee of experts on material safety. The book states that *"In the light of technical progress, the advent of new substances and materials, the exigencies of modern transport systems and, above all, the requirement to ensure the safety of people, property and the environment... the [international community needs revised] regulation of the transport of dangerous goods"* [45].

9.4.3 Stage 3: Formulators and Users

During Stage 3, the formulator or the user is at risk of exposure. However, the user may need to know their exposure to nanomaterials and the potential consequences. Exposure to nano-enabled products at this stage varies depending on the product's intended use. Using the product can lead to both intended and unintended consequences, such as inhalation, ingestion, ocular, and dermal exposure to engineered nanomaterials and their products.

Each industry has unique occupational exposure scenarios in product formulations and consumer exposure in final product use. Some products are ingested (as in medicines or foods), others are touched (as in electronics or toys), and others are inhaled (as in fragrances or cleaning products) [46–48]. In understanding nano-safety and risks in these exposure scenarios, specific consumer exposure information must be collected for high-production volume nanomaterials. This information is often collected by the United States Consumer Product Safety Commission (CPSC), an agency whose mission is to *"protect the public from unreasonable risks of injury or death associated with the use of the thousands of types of consumer products under the agency's jurisdiction"* (http://www.cpsc.gov). The specific consumer exposure information for one product can be applied to other consumer products when there are gaps in data. Simultaneously, hazard information is gathered and contextualized with exposure concentration (or dose) to characterize risk.

Within the occupational exposure of formulators, enclosed structures designed to prevent or reduce exposure to hazardous chemicals or vapors are just as crucial in this stage as in Stage 1, the production and manufacture [49]. It is important to remember that Stage 3 differs from Stage 1 in that Stage 1 exposure is often a single particle-type exposure (e. g., TiO_2 nanoparticles). In contrast, Stage 3 typically involves exposure to a mixture (e. g., TiO_2-enabled paint: nanoparticles plus latex, water, and pigment). In either stage, mechanical exhaust systems should be used to reduce contact with particles, vapors, or fumes. Some hazardous composite materials (e. g., polyvinyl chloride, arsenic, or benzene) are prohibited from aerosolization as they are very toxic to humans when inhaled. When working with nano-enabled products that are intentionally toxic to bacterial and viral organisms, such as cleaning products, integrated data sets of exposure information and material hazard data are needed to characterize the risks posed by these formulators or end-user exposures.

9.4.4 Stage 4: Disposal, Recycle, and Reuse

The nanomaterial's disposal, recycling, and reuse occur in the final life stage. Given the variety and variability among these processes, the risks associated with each nanomaterial must be considered on an individual product basis [50]. However, there are instances where products with many similar components or ingredients may be processed and analyzed together (i. e., plastics, electronics, metal scraps). When guiding workers to safely work with engineered nanomaterials in the end-of-life stage, determining if the material is included on any hazardous chemical watch lists is prioritized. As with any chemical disposal protocol, nanomaterials must be segregated by hazard class. Dry nanopowders should be packaged in cardboard boxes that weigh less than 10 kg. The container should be labeled with the contents, amount, date, and description for easy reference.

9.5 The Role of the Employer

In a nanomaterial manufacturing or formulation facility, the employer's role in promoting worker safety and awareness is vital to the company's success and its employees' health. Safe practices and adherence to Occupational Safety and Health Administration (OSHA) regulations start at the top of the corporate pyramid—with executive leadership—and descend to the personnel managers [51]. The job of senior leadership personnel requires developing and implementing an environmental safety committee devoted to the safe practice and education of nanotechnology. A safety committee or a risk management team could then maintain a safe working environment through three main strategies: safety information dissemination, safe practices training, and safety regulation enforcement.

Safety information dissemination is the first and most important defense against occupational accidents and injuries [52]. For workers to do their job safely and effectively, they must be well-informed about the engineered nanomaterials they are working with and trained on properly handling and disposing of the materials. Other safety precautions, such as PPE and standard operating procedures, must be followed to prevent exposure to nanomaterials, nano-waste, nano-intermediates, nano-enabled products, or other potentially harmful materials in the workplace. Safe practice training is a second safety measure utilized to prevent harm to the worker. This training aims to provide technical skills and hands-on experience in safe and appropriate practices in nano-enabled product development. Finally, enforcing safety regulations is necessary to ensure the workforce upholds the established protocols and standards every time the worker enters or exits the workplace. Together, the three strategies promote worker awareness and show confidence.

Creating a safe work environment is a continuous process that involves employers setting goals, providing feedback, and reinforcing good practices to improve safety. This iterative approach is like other risk management processes.

Risks associated with different nanoparticle states

Excerpt summarized from NIOSH document "General Safe Practices for Working with Engineering Nanomaterials in Research Laboratories"

- When nanomaterials are in a dry powder, handle with care so that it does not become airborne dust particles
- When in a liquid matrix, the risk for dermal exposure is high; there is also risk for aerosolization during certain procedures
- Nanomaterials in a solid matrix pose the least risk; if the matrix is disrupted (cutting, sawing, sanded, etc.), then the nanoparticles can be released

---

Administrative controls relevant to nanomaterial production

Excerpt summarized from NIOSH document "Current Strategies for Engineering Controls in Nanomaterial Production and Downstream Handling Processes
- Educate the workers on safe handling of nanomaterials
- Obtain a safety data sheets (SDS)
- Clean up spills in accordance with procedures
- Provide additional control measures (e. g., decontamination facilities, sticky mats, buffer area, restricted access to the lab)
- Conduct industrial hygiene and medical monitoring to ensure work practices are properly executed

---

9.6 Time to Reflect Questions

1. What is Behavior-Based Safety?
2. What are the six tasks completed by an occupational health officer?
3. Describe the four parts of a life cycle.
4. Why would you want to complete a risk assessment before you develop a nano-enabled product?
5. Explain an A-B-C analysis. What are the significant characteristics?
6. What are the benefits and limitations of conducting a behavioral assessment?
7. Explain how developing nano-enabled products differs from a traditional product.

Bibliography

[1] Piccinno F et al. Industrial production quantities and uses of ten engineered nanomaterials in Europe and the world. Journal of Nanoparticle Research, 2012, 14(9), 1–11.
[2] Cao G. Synthesis, properties and applications, World Scientific; 2004.
[3] Schulte P et al. Occupational risk management of engineered nanoparticles. Journal of Occupational and Environmental Hygiene, 2008, 5(4), 239–249.
[4] Warheit DB et al. Health effects related to nanoparticle exposures: environmental, health and safety considerations for assessing hazards and risks. Pharmacology & Therapeutics, 2008, 120(1), 35–42.
[5] Benson J. Safety considerations when handling metal powders. Journal of the Southern African Institute of Mining and Metallurgy, 2012, 112, 563–575.
[6] Oberdörster G. Pulmonary effects of inhaled ultrafine particles. International Archives of Occupational and Environmental Health, 2000, 74(1), 1–8.
[7] Wiesner MR et al. Assessing the risks of manufactured nanomaterials. Environmental Science & Technology. 2006, 40(14), 4336–4345.
[8] Oberdörster G et al. Principles for characterizing the potential human health effects from exposure to nanomaterials: elements of a screening strategy. Particle and Fibre Toxicology, 2005, 2(1), 1.
[9] Maynard AD. Nanotechnology: assessing the risks. Nano Today, 2006, 1(2), 22–33.
[10] Johnson DR et al. Potential for occupational exposure to engineered carbon-based nanomaterials in environmental laboratory studies. Environmental Health Perspectives, 2010, 49–54.
[11] Colvin VL. The potential environmental impact of engineered nanomaterials. Nature Biotechnology, 2003, 21(10), 1166–1170.

[12] Sotiriou GA, Diaz E, Long M, Godleski J, Brain J, Pratsinis SE, Demokritou P. A novel technique for toxicological characterization of engineered nanomaterials. In: Nanotechnology 2011: Electronics, Devices, Fabrication, MEMS, Fluidics and Computational-2011 NSTI Nanotechnology Conference and Expo, NSTI-Nanotech 2011 (pp. 525–528), 2011.

[13] Roco MC. Nanotechnology: convergence with modern biology and medicine. Current Opinion in Biotechnology, 2003, 14(3), 337–346.

[14] Cooper D. Behavioral safety: A framework for success, B-Safe Management Solutions, 2009.

[15] Kreyling WG, Semmler-Behnke M, Möller W. Health implications of nanoparticles. Journal of Nanoparticle Research, 2006, 8(5), 543–562.

[16] Li N, Xia T, Nel AE. The role of oxidative stress in ambient particulate matter-induced lung diseases and its implications in the toxicity of engineered nanoparticles. Free Radical Biology & Medicine, 2008, 44(9), 1689–1699.

[17] Danna K, Griffin RW. Health and well-being in the workplace: A review and synthesis of the literature. Journal of Management, 1999, 25(3), 357–384.

[18] Guldenmund FW. The nature of safety culture: a review of theory and research. Safety Science, 2000, 34(1), 215–257.

[19] Geller ES. Ten principles for achieving a total safety culture. Professional Safety, 1994, 39(9), 18.

[20] Schulte PA, Salamanca-Buentello F. Ethical and scientific issues of nanotechnology in the workplace. Ciência & Saúde Coletiva, 2007, 12(5), 1319–1332.

[21] Gurjar BR, Mohan M. Environmental risk analysis: problems and perspectives in different countries. RISK, 2002, 13, 1.

[22] Wijnhoven SW et al. Nano-silver–a review of available data and knowledge gaps in human and environmental risk assessment. Nanotoxicology, 2009, 3(2), 109–138.

[23] Blaser SA et al. Estimation of cumulative aquatic exposure and risk due to silver: contribution of nano-functionalized plastics and textiles. Science of the Total Environment, 2008, 390(2), 396–409.

[24] Cross SE et al. Human skin penetration of sunscreen nanoparticles: in-vitro assessment of a novel micronized zinc oxide formulation. Skin Pharmacology and Physiology, 2007, 20(3), 148–154.

[25] Chiou GC. Technology review: Toxic responses in the eye and visual system. Toxicology Methods, 1992, 2(3), 139–167.

[26] Yah CS, Simate GS, Iyuke SE. Nanoparticles toxicity and their routes of exposures. Pakistan Journal of Pharmaceutical Sciences, 2012, 25(2).

[27] Nurkiewicz TR et al. Nanoparticle inhalation augments particle-dependent systemic microvascular dysfunction. Particle and Fibre Toxicology, 2008, 5(1), 1.

[28] Gaillet S, Rouanet J-M. Silver nanoparticles: Their potential toxic effects after oral exposure and underlying mechanisms–A review. Food and Chemical Toxicology, 2015, 77, 58–63.

[29] Lyu Z et al. Developmental exposure to silver nanoparticles leads to long term gut dysbiosis and neurobehavioral alterations. Scientific Reports, 2021, 11(1), 6558.

[30] Köhler AR et al. Studying the potential release of carbon nanotubes throughout the application life cycle. Journal of Cleaner Production, 2008, 16(8), 927–937.

[31] Bauer C et al. Towards a framework for life cycle thinking in the assessment of nanotechnology. Journal of Cleaner Production, 2008, 16(8), 910–926.

[32] Som C et al. The importance of life cycle concepts for the development of safe nanoproducts. Toxicology, 2010, 269(2), 160–169.

[33] Beaudrie CE, Kandlikar M, Satterfield T. From cradle-to-grave at the nanoscale: gaps in US regulatory oversight along the nanomaterial life cycle. Environmental Science & Technology, 2013, 47(11), 5524–5534.

[34] Hischier R, Walser T. Life cycle assessment of engineered nanomaterials: state of the art and strategies to overcome existing gaps. Science of the Total Environment, 2012, 425, 271–282.

[35] Donaldson K et al. Combustion-derived nanoparticles: a review of their toxicology following inhalation exposure. Particle and Fibre Toxicology, 2005, 2(1), 10.

[36] Hoet PH, Brüske-Hohlfeld I, Salata OV. Nanoparticles–known and unknown health risks. Journal of Nanobiotechnology, 2004, 2(1), 12.

[37] Monazam ER, Shadle LJ, Shamsi A. Spontaneous combustion of char stockpiles. Energy & Fuels, 1998, 12(6), 1305–1312.

[38] Dimitrijevic D. Dangers of nanotechnology: potential fire concerns and safety frameworks. International Journal of Emergency Management, 2010, 7(3), 249–257.

[39] Koponen I, Jensen K, Schneider T. Sanding dust from nanoparticle-containing paints: physical characterisation. Journal of Physics. Conference Series, IOP Publishing, 2009.

[40] Schneider T, Jensen KA. Relevance of aerosol dynamics and dustiness for personal exposure to manufactured nanoparticles. Journal of Nanoparticle Research, 2009, 11(7), 1637–1650.

[41] Corn M. Overview of environmental evaluation and control. CHEST Journal, 1981, 79(4 Supplement), 91S–94S.

[42] Murashov V, Howard J. Risks to Health Care Workers from Nano-Enabled Medical Products. Journal of Occupational and Environmental Hygiene, 2015, 12(6), D75–D85.

[43] Laboratory, N.R.C.C.o.P.P.i.t., Evaluating Hazards and Assessing Risks in the Laboratory. 2011.

[44] Savolainen K et al. Nanosafety in Europe 2015–2025: Towards safe and sustainable nanomaterials and nanotechnology innovations. Helsinki: Finnish institute of occupational Health, 2013.

[45] (UNECE), U.N.E.C.f.E. Recommendations on the Transport of Dangerous Goods – Manual of Tests and Criteria, U. Nations, editor. United Nations Publication, New York and Geneva, 2015.

[46] Sahay G, Alakhova DY, Kabanov AV. Endocytosis of nanomedicines. Journal of Controlled Release, 2010, 145(3), 182–195.

[47] Hecht DS et al. Carbon-nanotube film on plastic as transparent electrode for resistive touch screens. Journal of the Society for Information Display, 2009, 17(11), 941–946.

[48] Li Q et al. Antimicrobial nanomaterials for water disinfection and microbial control: potential applications and implications. Water Research, 2008, 42(18), 4591–4602.

[49] Tsai S-JC et al. Airborne nanoparticle exposures associated with the manual handling of nanoalumina and nanosilver in fume hoods. Journal of Nanoparticle Research, 2009, 11(1), 147–161.

[50] Gao P et al. A simple recycling and reuse hydrothermal route to ZnO nanorod arrays, nanoribbon bundles, nanosheets, nanocubes and nanoparticles. Chemical Communications, 2009, 19, 2762–2764.

[51] Maakestad WJ, Helm C. Promoting Workplace Safety and Health in the Post-Regulatory Era: A Primer on Non-OSHA Legal Incentives That Influence Employer Decisions to Control Occupational Hazards. North Kentuck Law Review, 1989, 17, 9.

[52] Reason J. The contribution of latent human failures to the breakdown of complex systems. Philosophical Transactions of the Royal Society of London. Series B, Biological Sciences, 1990, 327(1241), 475–484.

Dominick Fazarro

10 The Future of Nanotechnology Safety

Before moving forward with this chapter, the editors would like to tell a fictional story of things to come in the future. If anyone, from top management down to the worker, neglects the Occupational Safety and Health Administration (OSHA) standards, the normal protocol is that OSHA conducts an investigation of a potential violation reported by a worker(s). If a violation is found, then the company receives a monetary fine and an abatement plan. The situation is no different for the manufacturing and handling of nanomaterials.

On 20 April 2035, OSHA receives a complaint from an undisclosed company that there has been an incident of spillage of a highly advanced nanomaterial used for coatings for aviation purposes. The complaint also indicates that management does not have the proper personal protective equipment (PPE), ventilation systems, or adequate training to safely handle and work with these highly advanced nanomaterials. As a result, the company is fined. With new standards and abatement programs in place for nanomaterials, OSHA can effectively do their job based on years of research of hazards, engineering and administration controls, toxicity findings, and advances in PPE equipment. The research findings were crucial in establishing standards and protocols for worker safety. However, even in the year 2035, we as observers and industrialists cannot escape human fallibility.

The purpose of this book is to assist manufacturers and managers understand the dangers and hazards of handling nanomaterials. This is currently one of a few books discussing nanotechnology safety, which is an indication that researchers are becoming concerned. As you may know, technology acts as a double-edged sword, meaning there are good and bad aspects. Humans are fallible and make mistakes. According to an article written by Behavioral Safety Management Systems (BSMS, 2017) [1], who are experts in behavioral-based safety, "People often behave unsafely because they have never been hurt before while doing their job in an unsafe way: 'I've always done the job this way' being a familiar comment."

Nanotechnology, compared with microlevel technology, is a field where the safety required is ten times the current requirement for human attention and requires a lower threshold of human error to prevent hazards. This is not say that people cannot work under a comprehensive safety environment with nanomaterials, but the level of knowledge to identify hazards and to act to control/prevent hazards requires a high level of understanding and action in the face of uncertainty.

Engineers and scientists have gathered information on over 650,000 known chemicals used in industry. For decades, data on toxicity, associated hazards, physical and chemical properties, hazard identification, exposure controls, and PPE have been collected to help workers understand the careful handling of dangerous materials. Now science has entered the 1–100 nm realm, where the majority of safety measures used

https://doi.org/10.1515/9783110781830-010

for chemicals are virtually useless, because (i) the properties of these new materials are unknown; (ii) properties change according to nano-size; (iii) there are no data on the effects of exposure for either humans or the environment; (iv) current available equipment cannot operate effectively at the nanoscale; and (v) guidance and regulations are contradictory. The list is actually much larger, but these are the main points.

On a positive note, there is a promising side to the longevity of nanotechnology. Let us take a look into the future of nanotechnology safety. In the year 2035, nanomaterials and other advanced materials will be commonplace for enhancing technology for consumer and commercial goods. The quality of life for humans will have increased due to advanced medicines using nanomaterials. Federal safety regulations and standards will be fully adopted by industries. Advanced safety controls operated by artificial intelligence will be able to decrease human error to almost zero. This is all very possible, but will take national and international efforts to make it a reality. In the world of higher education, nanotechnology safety will be its own discipline, because use of nanomaterials will be commonplace and a necessity for industry. To work in an area of nanomaterials, certification will be required for professionals.

The future outlook is hopeful if we implement appropriate controls for the application of innovative technology for production and use of nanomaterials in areas such as consumer goods, medicine, and the military. The world we live in is based on remaining profitable in order to remain in business, even though scientists developing nanomaterials see them as a new industrial revolution or the next big step resulting from human ingenuity.

Industrialized nations can embrace the unlimited possibilities of nanotechnology; however, our paradigm of thinking and how we approach technologies must change. National and international governing bodies need more stringent guidelines for worker safety regarding nanomaterials and not guidelines based on political correctness. As we progress into the future with highly advanced nanomaterials, safety protocols must evolve to stay current.

Even though nanotechnology is relatively new in society, the unknowns keep safety managers up all night trying to find ways to maintain a high level of safety. The authors believe that more government funding is needed for research into the hazards of current and new nanomaterials and for designing new PPE and engineering controls. As more nanomaterials are created, industry must be cognizant of constant improvement in the training needs of workers. A workforce well-trained in safely handling nanomaterials will lessen the likelihood of catastrophes and decrease public skepticism.

Bibliography

[1] BSMS. Psychology of behavioral safety, 2017. Retrieved from http://www.behavioral-safety.com/component/content/article/3-psychology/2-the-psychology-of-behavioral-safety.

Index

https://doi.org/10.1515/9783110781830-011

www.ingramcontent.com/pod-product-compliance
Lightning Source LLC
Chambersburg PA
CBHW061416210326
41598CB00035B/6231